Earth's Deep History

Published with support of the Susan E. Abrams Fund

# EARTH'S
# Deep History

How It Was Discovered
and Why It Matters

MARTIN J. S. RUDWICK

THE UNIVERSITY OF CHICAGO PRESS

CHICAGO AND LONDON

The University of Chicago Press, Chicago 60637
The University of Chicago Press, Ltd., London
© 2014 by The University of Chicago
All rights reserved. Published 2014.
Paperback edition 2016
Printed in the United States of America

25  24  23  22  21  20  19  18  17  16      4  5  6  7  8

ISBN-13: 978-0-226-20393-5        (cloth)
ISBN-13: 978-0-226-42197-1        (paper)
ISBN-13: 978-0-226-20409-3        (e-book)
DOI: 10.7208/chicago/9780226204093.001.0001

Library of Congress Cataloging-in-Publication Data

Rudwick, M. J. S., author.
    Earth's deep history : how it was discovered and why it matters /
Martin J. S. Rudwick.
        pages cm
    Includes bibliographical references and index.
    ISBN 978-0-226-20393-5 (cloth : alkaline paper) —
ISBN 978-0-226-20409-3 (e-book)    1. Earth sciences—History.    2. Natural
history—History.    3. Religion and science.    I. Title.
    QE11.R827 2014
    550—dc23
                                                            2014010242

For Trish

*Deo gratias*

CONTENTS

# INTRODUCTION

Sigmund Freud once claimed that three great revolutions had transformed our human sense of our place in nature. The first had removed our Earth from the center of the universe, turning it into one planet among several others, orbiting one ordinary star among a vast multitude of others. The second revolution had embedded our species in the rest of the animal world, by supposedly demoting us from being the objects of God's unique concern and turning us into mere naked apes. And the third revolution had undermined any sense of ourselves as rational beings, by disclosing the depths of our unconscious fantasies. These major changes in our conception of ourselves have subsequently been labeled with celebrated names, respectively those of Copernicus, Darwin, and Freud himself.

However, as my friend the late Stephen Jay Gould pointed out long ago, Freud's list omitted a fourth revolution that certainly deserves a place in the same league, although it lacks the convenience of being associated with any single well-known individual. One striking feature of this fourth great change—the second in historical order—was that it vastly enlarged the timescale of our Earth and by implication that of the universe, just as the first or Copernican revolution had vastly enlarged its spatial scale. In earlier times, most people in the West had taken it for granted that the world had started, if not precisely in 4004 BC, then at some such point in time, only a few millennia ago. After this revolution it became equally commonplace to accept that the Earth's timescale runs at least into millions of years, if not billions. Geologists now work routinely with mind-boggling amounts of "deep time," just as their colleagues the astronomers and cosmologists work with literally inconceivable magnitudes of cosmic "deep space" (and time too).

This much is now well known, far beyond scientific circles. But an overwhelming emphasis on the enlargement of the timescale has obscured two other features of this great revolution, which, taken together, are much more significant. The first of these was a radical change in the place of humanity itself. The "young Earth" of the traditional picture was also an almost wholly human Earth. Apart from a brief opening scene or prelude—putting the props on stage, as it were—it was a human drama from start to finish, from Adam through to some future Apocalypse at the end of the world. In contrast, the "ancient Earth" first discovered and reconstructed by early geologists was largely non-human because it was almost completely pre-human: our species seemed to have made a very late appearance on the world stage. Most of this newly discovered deep time was therefore as devoid of any human presence as the vastnesses of deep space.

At the same time, the distinction between a relatively brief human period and a far more lengthy pre-human one was a sign of a second and even more radical consequence of this great revolution in our conception of nature. The simple sequence of a non-human period followed by a human period was enough in itself to give our planet a basically *historical* character; and the vast expanses of pre-human deep time, even on their own, turned out to have been filled with a history just as eventful and dramatic in its own way as human history. In short, it turned out that *nature has had a history of its own*.

So this book offers a brief account, not primarily of the discovery of deep time, but rather of the reconstruction of *Earth's deep history* and our human place within it. The story of this fourth great revolution has been neglected, particularly in books and TV programs designed for the general public. There are two distinct reasons for this. First, it has been shrunk into nothing more than a prelude to the supposedly more exciting story of Darwin's theory of evolution. It is true that the recognition of the Earth's deep history was a necessary precondition for any satisfactory explanation of the diversity of living organisms, and particularly of the origin of our own species. But the story summarized in this book has had a career of its own, independent of Darwin's or any other theory of evolution, because it concerns the history of *everything* on Earth: not only plants and animals, but also rocks and minerals; mountains, volcanoes, and earthquakes; continents, oceans, and atmosphere. So the recognition that the Earth has had a history of its own, and that it was possible to reconstruct it reliably and often in detail, amounted to a major revolution in human thought. It is a story that deserves to be told in its own terms and for its own sake.

The second reason for this story's neglect is that it has been shrunk into just one episode in the triumphant march of Science in its struggle against Religion. The notorious 4004 BC date already mentioned has been widely taken to typify the repressive obscurantism of The Church in resisting the progress of Enlightened Reason. But this use of labels such as Science and Religion, Church and Reason (usually in the singular and often with initial capitals), should make us suspicious. Real history is never so abstract or so tidy. In fact, this stereotype of a perennial conflict between Science and Religion has long been abandoned by historians who have studied any of its alleged episodes at all closely. It makes shoddy history, though of course it provides stirring rhetoric for modern atheistic fundamentalists. In this book, in contrast, I try to show how an emerging sense of the Earth's deep history was related to earlier conceptions of a much briefer kind of history in far more interesting and important ways than this tired stereotype allows. The surprising revival of "young Earth" ideas by some modern religious fundamentalists, and the even more surprising political power of such ideas in certain parts of the world, should not distract us from tracing the main story. I deal briefly with the modern creationists at the very end of this book, but in such a way that I hope it will be clear that they are a bizarre sideshow, not the climax of the narrative.

I argue in fact that the discredited stereotype of perennial conflict between Science and Religion should, at least in this case, be turned upside down. Once we recognize that the core of this great revolution in human thought lay in the realization that nature has had a history of its own, the merely quantitative enlargement of its timescale becomes a secondary issue. What is much more important is to understand the origins of this new sense of the historicalness or *historicity* of nature. It should be no surprise that its source lay in the contemporary understanding of *human* history, which was deliberately and knowingly transposed into the world of nature. Human history, not physics or astronomy, became the model for tracing the history of nature. The rise and fall of empires, for example, was utterly unpredictable even in retrospect, unlike the predictable movements of the planets. Human history was recognized as being deeply *contingent*: at every point things could well have turned out differently (this alone makes it possible, and often fascinating, to ask counter-factual or "what if . . .?" questions about the past). This was the sense of historicity that was transferred from culture into nature, generating a new understanding of nature, and specifically of the Earth, as similarly historical. If this transfer does seem surprising, it is probably because it entails accepting that the sciences of nature have here been decisively enriched by an input from the sciences of human history, right across the supposed gulf between the so-called Two Cultures, between Science and the humanities. People outside the English-speaking world don't experience the same difficulty, because they have the good sense to call *all* these bodies of disciplined knowledge "sciences," in place of our peculiar Anglophone use of a singular "Science" for just some of them.

In view of the character of Western culture during the relevant centuries (roughly, seventeenth through nineteenth), it should also be no surprise that one major source—even arguably *the* major source—for this new vision of nature as historical was the strong sense of history embodied in the Judeo-Christian scriptures, with their dynamic narrative thrust from primal Creation through pivotal Incarnation towards an ultimate City of God. These culturally foundational texts, far from obstructing the discovery of the Earth's deep history, positively facilitated it. To borrow a metaphor from biology, they *pre-adapted* their readers to find it easy and congenial to think in similarly historical terms about the *natural* world that formed the context of human action and, so believers claimed, of divine initiative. Of course this suggestion is neutral with respect to the validity of the religious perspective embodied in the texts: it does not amount to evidence in favor of these religious beliefs or against them, and my purpose in making the connection is historical, not apologetic.

Does the discovery of the Earth's deep history matter? Certainly it is a fascinating story in its own right, and one that deserves to be far more widely known: contrast its low profile with the huge attention that was rightly given to Darwin's evolutionary theory in his bicentennial year. Beyond its intrinsic interest, I believe it matters profoundly, because it disclosed something about our world that has wide-ranging implications and was quite unexpected. Those who in earlier times made it their business or their vocation to study the world of nature—the people who have come to be called scientists—widely assumed that with further study it would become more and more *predictable*. They aimed to uncover the "laws" of nature, which by definition were taken to be the same yesterday, today, and forever. The better the laws of nature were understood, the more effectively human individuals and societies would be able to control or change the world of nature in the service of human goals and purposes. Sciences such as physics and astronomy were therefore taken as models. The more the underlying laws of nature were quantified and given mathematical expression, the more precisely the timing of an eclipse, for example, could be predicted.

In contrast, the discoveries outlined in this book showed that the Earth's deep history—and therefore its future—could not be reduced to any such simple and predictable form. The Earth had not been programmed, as it were, in such a way that its past and future course was fully determined, given certain initial conditions and the unchanging laws of nature. Of course the component parts of terrestrial nature were assumed to be acting indeed according to unchanging laws: the power of crashing waves to erode a coastal cliff, for example, was taken to have been underlain in the deep past by the same laws of physics as at the present day. But the past history and likely future of *this* continent and *this* ocean could not be deduced from any such nonhistorical laws, still less the past and future of the Earth as a whole. All such histories had to be reconstructed from surviving evidence of what *in fact* had happened, just as the past history of the people inhabiting the land and trading across the sea had to be reconstructed from the surviving documents and artifacts of their history. In other words, the Earth's deep history could not be reconstructed by applying the laws of nature "top down," but only by piecing together the historical evidence "bottom up." The Earth's deep history turned out to have shared the messy unpredictable contingency of human history, rather than the astonishingly precise predictability of, say, the motions of the Moon and planets in relation to the Sun. That this unpredictable contingency *matters*—not least in current controversies about our human role in the near future of our home planet—should need no further emphasis.

In the course of human history, the science of geology was the first to develop this new sense of nature itself as intrinsically historical, but it was not the last or the only such science. Just as geologists came to recognize that, say, the present form of the Alps cannot be understood without unraveling the long and complex *history* of those mountains, so biologists—and notably Darwin who, significantly, began his career as a geologist—later showed that the present forms and habits of plants and animals likewise embody their own evolutionary histories and cannot be fully understood without taking those histories into account. And the same kind of historicity was eventually adopted even in the largest-scale science of all: cosmologists now deal routinely with the reconstruction of the *histories* of stars and galaxies—and even the history of the entire universe from its conjectural Big Bang onwards—in ways that are closely parallel to those first developed by geologists for the Earth's deep history. So the story I summarize in this book has an importance that goes far beyond the particular science on which it is focused.

In conclusion, I must emphasize that this book is based, as any such work should be, not only on my own historical research but also on research by many other historians of many nationalities, most of it published in recent decades and in several languages. This needs to be emphasized, because all this modern research by historians of the sciences is too often blithely ignored, or at best under-utilized—with a few honorable exceptions—by the authors of popular science books, by the makers of TV science programs, and, most seriously, by scientists who pronounce on the history of their own sciences. They all seem to prefer to stay in a cozy comfort zone of recycled myths about the past, often myths with an unattractively chauvinistic (and sexist) flavor, singling out "The Father" of this or that.

In view of the sheer mass of reliable historical research that is available, the writing of this short book has demanded a drastic pruning of detail, and a sharpening of focus, in order to highlight what I see as the main features of the story. In particular, I have concentrated this account on the arguments and activities of those who came to call themselves scientists, rather than the ideas that were prevalent among wider social groups or in society as a whole. I have touched only lightly on the broader cultural implications of what these people claimed to have discovered. And it is a matter of human history that most of the basic ideas about our planet's deep history, which now underlie the work of Earth scientists worldwide, were first developed in Europe and not elsewhere. So most of my story is focused on the European cultural sphere rather than those other parts of the world that play an increasingly important role in the sciences of the 21st century. (If the story is also largely one of male activities, that is because it reflects earlier historical realities; a more detailed history of the last few decades would show that, at least in this kind of science, gender has become increasingly irrelevant.)

I hope this book will help not only to make a great revolution in human thought more widely known and understood, but also to blow away the cobwebs of some outdated ideas: not least the pervasive myth of perennial conflict between "Science" and "Religion," two beasts as mythical in every sense as those traditional symbols of good and evil, St. George and the Dragon.

# 1

# Making History a Science

"Time we may comprehend: 'tis but five days elder than ourselves." So the 17th-century English writer Sir Thomas Browne summarized, almost casually, the profound question of the ultimate origin of our world, our species and time itself. In the age of scientific giants such as Galileo and Newton, most people in the Western world, whether religious or not, took it for granted that humanity is of almost the same age as the Earth. They also assumed that not just the Earth, but the whole universe or cosmos, and even time itself, are scarcely any older than human life.

The opening chapter of Genesis, and of the Bible, set out a brief narrative in which Adam ("The Man") had been formed on the sixth day of creative action, after five days of preparation and before God completed a primal week by resting on its Sabbath day. Browne and his contemporaries did not need a repressive Church to bully them into accepting this as a reliable account of the most distant past (and anyway, in a Christendom fractured by the Reformation and Counter-Reformation, there was no single all-powerful body capable of enforcing any such belief). It seemed obvious *common sense* to them that the world must always have been a *human* world, apart from a brief prelude in which the props necessary for human life had been put on stage: Sun and Moon, day and night, land and sea, plants and animals. A world without human beings would have struck them as utterly pointless, except as a brief setting of the scene for the human drama to come. So they took it for granted that Genesis gave them an authentic account of the world's earliest origins. It came, they believed, from the hand of Moses, the only ancient historian to have recorded the earliest ages of the world; and the very first phase of that history—before any human being had been there to witness and remember it—could only have been disclosed to Moses (or to Adam before him) by the Creator himself. To cap it all, nothing in the world around them seemed obviously to suggest that its history had been otherwise.

Browne and most of his contemporaries, educated and uneducated alike, took it for granted that the history of humanity was of almost the same length as the history of the natural world. But far from thinking these histories were very short, and the Earth very young, they regarded both as extremely long, relative to brief human lives of, at best, some "three score years and ten." History was plotted on a scale of the "Years of the Lord" (*Anni Domini*, AD) that had elapsed since Jesus's birth, which was treated as the uniquely pivotal moment of divine Incarnation. Since that point in time and the time, some thirty years later, when the Roman official Pontius Pilate had ordered Jesus's execution, more than sixteen centuries had passed into history. This was a very long span of time by any human standard; the study of the Romans and their highly respected Latin literature fully deserved its title of "*Ancient* History." Yet the scale of "Years Before Christ" (BC) stretched even further back, past the ancient Greeks and their equally admired literature, to the obscure earliest ages for which the only surviving records were widely believed to be those in the Bible. Most historians reckoned that the primal Creation itself must be nearly three times as distant from the Incarnation as the Incarnation was distant from their own day. In total this amounted to an almost inconceivably lengthy history of the world. Some fifty or sixty *centuries* seemed more than enough time for the unfolding of the whole of known human history and also therefore for the natural world, the stage on which it had been played out. The world's beginnings put even the "Ancient History" of the Greeks and Romans into the shade.

When one of these 17th-century historians calculated that the week of Creation had started on a specific day during the year 4004 BC, the date could be questioned, and was, but the precision aimed at was not. Nor was the order of magnitude thought to be an underestimate. This particular figure was published by James Ussher, an Irish historian whose powerful patron and great admirer had been King James I of England (James VI of Scotland). Shortly before that monarch's death, he appointed Ussher to be Archbishop of Armagh and head of the established Protestant church in Ireland, though as it happened the scholar spent most of his later life in England.

In modern times, Ussher and his date of 4004 BC have been much scorned and ridiculed. But Ussher was not a religious fundamentalist in the modern mold. He was a public intellectual in the mainstream of the cultural life of his time. His work doesn't deserve to be treated as a joke like those in *1066 And All That*, the classic spoof history in which the English national story is studded with unmistakeable Good Kings and Bad Kings, Good Things and Bad Things. Ussher's 4004 BC was not, in its time, a Bad Thing. On the contrary, what it represented was in some important respects a thoroughly Good Thing. Ussher's view of world history may seem so far removed from the modern scientific picture of the Earth's deep history that there can be no possible link between them, except as irreconcilable alternatives (which, in the eyes of modern fundamentalists, both religious and atheistic, is just what they are). In fact, however, what 17th-century historians such as Ussher were doing is connected without a break with what Earth scientists are doing in the modern world. Ussher is therefore a good starting point for understanding the origins of our modern conception of the Earth's deep history. Moreover, once Ussher's ideas are understood in the context of his own time, their superficial similarity to modern creationist ideas of a "Young Earth" is transformed into a stark contrast. The creationists, unlike Ussher, are out on a limb, and a precarious one at that.

In the 17th century Ussher was just one of the many scholars, scattered across Europe, who were engaged in the kind of historical research that was called "*chronology*." This was an attempt to construct a detailed and accurate timeline of world history, compiled from all available textual records, both sacred and secular, including records of striking natural events such as eclipses, comets, and "new stars" (supernovae). Other chronologists criticized or rejected many specific details in Ussher's timeline, but most of them shared his broader aims, and his compilation illustrates very well what they were all trying to do.

Ussher published his *Annals of the Old Covenant* (*Annales Veteris Testamenti*, 1650–54) near the end of a long and highly productive scholarly life. He wrote it in Latin, which ensured that it could be read by other scholars elsewhere: Latin was the common international language of educated people throughout Europe, just as English is today around the world. Ussher's two massive volumes were entitled *Annals* because they summarized year by year what was known of events in world history; or at least he assigned each event to what he judged to be its correct year, and described them all in strict temporal order. So his book began with Creation at 4004 BC. But it extended forwards right through the BC/AD divide and the years of Jesus's life, as far as the immediate aftermath of the Romans' utter destruction of the great Jewish Temple in Jerusalem in AD 70. From Ussher's Christian perspective, this marked the decisive end of the "Old Covenant" linking God specifically with the Jewish people. So his chronology traced the course of world history as far as the first few years of God's "New Covenant" with the new people of God—in principle global and multiethnic—represented by the Christian Church.

Ussher's world history embodied the best scholarly practice of his time. Chronology fully deserved its status as a historical *science* (using that word in its original sense, which is still current except in the Anglophone or English-speaking world). It was based on a rigorous analysis of all the ancient textual records known to him. These were mostly derived from sources in Latin, Greek, and Hebrew. Half a century earlier, the French scholar Joseph Scaliger, the greatest and most erudite chronologist of them all, had also used those in several other relevant languages such as Syriac and Arabic. But even Scaliger knew only a little about sources further afield, for example from China or India, and the ancient Egyptian hieroglyphs had not yet been deciphered. Nonetheless, chronologists had available to them a massive body of multicultural and multilingual evidence. From all these varied records they extracted dates such as those of major political changes, the reigns of ancient monarchs, and memorable astronomical events. They then tried to match them up, often across different ancient

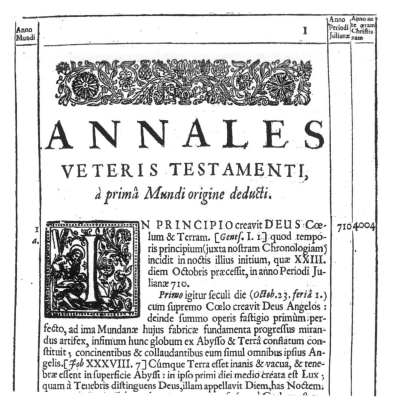

**FIG. 1.1** How Ussher's "4004 BC" first appeared in print: part of the opening page of his *Annals of the Old Covenant* (1650–54), with his dating system in three marginal columns. On the left, the "Year of the World" [*Anno Mundi*] starts at 1, with Creation itself. On the right, the "Year before the Christian Era" [*Anno ante æram Christianam*] starts at 4004, and will decline as the chronology proceeds, but the "Year of the Julian Period" [*Anno Periodi Julianæ*]—a kind of reference timeline independent of any real history—is already at 710. In the opening sentence of his text, Ussher dates the initial act of Creation, including the beginning of real time, at the start of the night preceding 23 October in the year 710 Julian, so that still earlier Julian years were in a kind of "virtual" time. Chronology was not a science for the simple-minded! Ussher's Latin title refers to the theological concept of a divine "old *covenant*" with the Jewish people, not to the Jewish scriptures or "Old *Testament*"; his chronology also covers the period to which the New Testament or Christian scriptures refer.

cultures, and to link them together in a continuous chain of dated events. (The science of chronology is not extinct: the results of modern chronological research are on display in our museums, wherever artifacts from ancient China or Egypt, for example, are labeled with dates BC or BCE; all such dates are derived from similar correlations between the histories of different cultures.)

By far the greater part of Ussher's evidence, like that of other chronologists, came not from the Bible but from ancient *secular* records. Not surprisingly, his sources were most abundant for the more recent centuries BC, and tailed off rapidly as he penetrated into the more remote past. For the very earliest times they were extremely scanty and almost confined to the bare record in Genesis of "who begat whom" in the earliest generations of human life. This makes it clear that Ussher's main objective was indeed to compile a detailed history of the world, and not primarily to establish the date of Creation or to bolster the authority of the Bible in general. Ussher treated the Bible as one historical source among many, even if it was also, from his perspective, the most valuable and reliable of all.

## DATING WORLD HISTORY

Like other chronologists, Ussher adopted the sophisticated dating system that had been devised by Scaliger. The Frenchman had constructed a deliberately artificial "*Julian*" timescale from astronomical and calendrical elements. It provided a neutral dimension of time, as it were, on which rival chronologies could be set out and compared. It was not just a convenient device; it also highlighted the crucial distinction between *time* and *history*. Time itself was just an abstract dimension measured in years; history was all the real events that had happened in the course of time. What any chronologist claimed as real history could be plotted, on a baseline of the Julian scale, as "years of the world" (*Anni Mundi*, AM) counting forwards from Creation, or as "years before Christ" (BC) counting backwards from the Incarnation, from which the "Years of the Lord" (AD) were counted forwards. Research on chronology was powered by an intellectual craving for quantitative precision. This was characteristic of the age, and not confined to projects such as chronology. It was even more prominent in the natural sciences, for example in the contemporary work of astronomers such as Tycho Brahe and Johannes Kepler. In both kinds of investigation, quantitative precision was valued more highly than ever before.

Like cosmology, however, chronology was a highly controversial kind of research. Producing a dated timeline of events was fraught with problems of incomplete, ambiguous, or incompatible records. At one point after another, chronologists had to use their scholarly

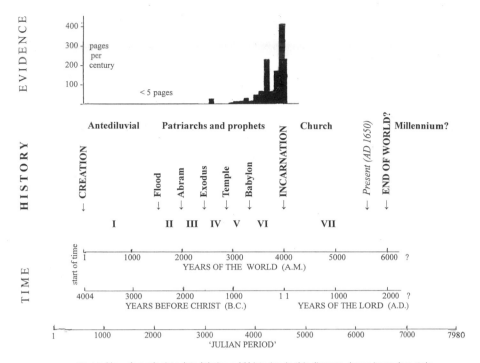

**FIG. 1.2** How chronologists dated their world histories. In this diagram, drawn in modern style, time flows from left to right. The "Julian period" was a deliberately artificial timeline, on which each of 7980 years, past and future, could be defined uniquely by a combination of astronomical and calendrical factors. It served as a reference scale of *time*, on which chronologists could plot the dates in *history* that they calculated for Creation, Noah's Flood, the birth of Christ, and other decisive events or "epochs," expressed either in years BC and AD, or in "Years of the World" (*Anni Mundi*, AM) from Creation. These events then defined seven "Ages" (I to VII) for the whole of world history, as seen of course from a Judeo-Christian perspective. This diagram is based on the figures in Ussher's *Annals*, but those calculated by other chronologists were not (on this scale) substantially different. The bulk of relevant historical records declined rapidly as chronologists penetrated back in time: the histogram shows the amount of text devoted, in Ussher's work, to successive centuries: his *Annals* started at 4004 BC and ended at AD 73.

judgment to decide which records were the most reliable, and how they could most plausibly be linked together in an unbroken timeline. Consequently, there were almost as many rival dates for each important event as there were chronologists proposing them. This was particularly true for the date of Creation itself. Ussher's 4004 BC was just one proposal in a crowded field ranging (according to one survey) from 4103 BC to 3928 BC. Scaliger, for example, had decided on 3949 BC, and Isaac Newton—a keen chronologist among many other

things—later settled for 3988 BC. Ussher, like some other chronologists though not all, claimed a very precise date indeed, namely the
start (at nightfall, according to Jewish timekeeping) of the first day of
the first week after the autumn equinox; this marked the Jewish New
Year equivalent to the Christian year 4004 BC. At the time, complex
calendrical and historical reasoning made this kind of precision a perfectly respectable ambition, however bizarre it may seem to us.

It is only by historical accident that Ussher's 4004 BC has become
the best known of all such dates and now the most notorious, at least in
the English-speaking world. Almost half a century after Ussher's death

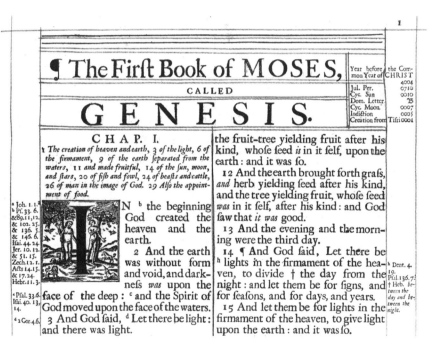

**FIG. 1.3** How Ussher's "4004 BC" first appeared in the Bible itself: part of the opening page of
William Lloyd's edition (1701) of the "Authorized" [King James] translation in English, with the
first or *hexameral* (six-day) Creation story at the start of the book Genesis. This page shows
inconspicuously (top right) the Creation dated at 4004 Before Christ, 0710 on the Julian scale,
0001 for the world itself, and other calendrical data. In the same marginal column, and also on
the left, are the first of many hundreds of editorial cross-references to other parts of the Bible,
and notes on the Greek and Hebrew texts on which the translation was based. This should have
have made it clear to readers—but often didn't—that the dates in the margins were likewise
editorial notes, not a part of the sacred text itself. The little picture decorating the first letter of
the text is of Adam and Eve in the Garden of Eden, the second Creation story in Genesis.

a scholarly English bishop included a long string of Ussher's dates among his own editorial notes in the margins of his new edition of the "Authorized" or "King James" translation of the Bible into English, which had originally been published with the authority of Ussher's royal patron back in 1611. Ussher's dates remained there, by custom or inertia, in successive editions of the Bible in English, right through the 18th century and most of the 19th, although they were never formally authorized by either church or state. Darwin and his English contemporaries, for example, would have grown up seeing 4004 BC printed on the very first page of their family Bibles. Many young or uneducated readers, not understanding the role of an editor, assumed that the date was an integral part of the sacred text, and they respected or even revered it accordingly. Only in 1885 were all Ussher's dates—by then long obsolete, in historical as well as scientific terms—omitted from the margins of the new "Revised Version" of the Bible. This was the first complete English translation to incorporate the greatly improved linguistic and historical understanding of the texts that was the fruit of biblical research by Jewish and Christian scholars since the time of Ussher (and King James). Readers of the Bibles placed by the Gideons in hotel bedrooms had to wait even longer, until the late 20th century, to be relieved of the implications of 4004 BC. In contrast, marginal dates did not usually feature in Bibles in other languages, so people outside the English-speaking world were generally spared this disastrous misapprehension that the exact date of primal Creation had been fixed by divine, or at least ecclesiastical, authority.

## PERIODS OF WORLD HISTORY

To return, however, to Ussher's century: his and other chronologists' efforts to compile rigorously precise "annals" of world history were a means to what most of them regarded as a more important end. Quantitative precision was intended to help yield qualitative meaning. Chronologists wanted to give precision to what they saw as the overall shape of human history, by dividing it into a meaningful sequence of periods. The primary division represented by the traditional dating system of years BC and AD was just such a distinction, for it separated the old human world before the Incarnation from the radically new human world which—from a Christian perspective—that unique

## ÆTAS MVNDI SECVNDA.

| 1657 a. | **A**Nno fexcentefimo primo vitæ Noachi, menfis primi die primo, (*Octob.*23. *feriâ* 6. ut *novi Anni* ita & novi *Mundi* die primo) cùm ficcata effet fuperficies terræ, removit Noachus operculum Arcæ. [*Genef.*VIII.13.] <br> Menfis 2 die 27. (*Decemb.* 18. *feriâ* 6.) cùm exaruiffet terra, Dei mandato exivit Noachus, cum omnibus qui cum ipfo fuerant in Arca. [*c*. VIII.14,-19.] <br> Noachus egreffus Soteria Deo immolavit. Deus rerum naturam, diluvio corruptam, reftauravit : carnis efum hominibus conceffit ; atque Iridem dedit fignum foederis. [*c*.VIII.& IX.] <br> *Anni vitæ humanæ quafi dimidio breviores fiunt.* | | |
|---|---|---|
| 1658 d. | Arphaxad natus eft Semo centenario, biennio poft diluvium, [*c*.XI. 10.] finitum fc. | 2368 | 23.46 |
| 1693 d. | Salah natus eft; quum Arphaxad pater ejus 35 vixifset annos. [*c*.XI. 12.] | 2403 | 2311 |
| 1723 d. | Heberus natus eft ; quum Salah Pater ejus 30 vixifset annos. [*c*.XI. 14.]                 Quum | 2433 | 2281 |

**FIG. 1.4** The very early part of Ussher's *Annals* that covers Noah's Flood, with references to the Genesis text, which he believed was the only reliable source available at this remote point in world history. The Flood is dated at Years-of-the-World 1657 (marginal column on left, as in Fig. 1.1); the first postdiluvial generations are then recorded with, in addition, Julian and BC dates (in columns on right, again as in Fig. 1.1). For Ussher and other chronologists it was also of great importance that the Flood marked the start of the world's "Second Age" [*Aetas Mundi Secunda*]. Apart from the initial six "days" of Creation, the Flood was the event in scriptural history that most obviously involved the natural world as well as the human. This made it the focus of much subsequent debate about how early human history might be related to the Earth's own history.

event had first brought into being. But Ussher, like other chronologists, also subdivided the millennia of BC history, by defining a sequence of decisive events or *"epochs,"* which in turn marked out a sequence of distinctive *"ages," "eras,"* or periods. Ussher identified five significant turning-points between the mega-events of the Creation and the Incarnation. These ranged in time from Noah's Flood to the ancient Jews' deportation into exile in Babylon. Adding the period since the Incarnation, world history could then be divided into a sequence of seven ages. These were often taken to match, or echo symbolically, the sequence of seven "days" in the week of Creation itself. So the whole shape of world history was deeply imbued with Christian meaning.

In the 17th century, then, world history was pictured qualitatively

as a sequence of distinctive periods bounded by particularly significant events, each of which chronologists tried to date accurately on a
quantitative timescale. All this history was taken to be, most importantly, one of cumulative divine self-disclosure or *"revelation,"* but it
was also largely *human* history. The non-human world of nature was
treated for the most part just as a setting for the human drama, an almost unchanging background or context for human action and divine
initiative. Only occasionally did events in the natural world feature
prominently in accounts of human history, either sacred or secular. In
the sacred story, for example, the waters of the Red Sea had retreated
temporarily, enabling the Jewish people under Moses' leadership to
make their Exodus from Egypt and gain their freedom. Equally conveniently, or providentially, the Sun "stood still" for an embattled Joshua
(though what exactly that meant was much debated); later still, Jesus's
birth and death were said to have been marked by, respectively, a new
star and an earthquake.

Only at two points did the natural world feature still more prominently, right in the foreground of the sacred story. These two points
were the Creation itself, and Noah's Flood. In the 17th century, each
was the focus of a distinctive kind of historical commentary, which
to a limited extent enlarged the scholarly study of texts with materials
drawn from nature.

The first kind of commentary was on the six "days" or phases of
Creation. The brief narrative in Genesis was often used as a framework for reviewing what was currently known about the structure
and functions of the cosmos, the Earth, and plants and animals, all of
which jointly constituted the environment of human life. These commentaries (known as *"hexahemeral"* or *"hexameral,"* from the Greek
for "six days") followed what was taken to be the primary meaning of
the biblical text. They treated the origins of the major features of the
natural world as a coherent sequence of historical events in real time.
The narrative was taken to be describing the sequence in which the
props had been placed on stage, as it were, before the human drama
could begin. So any such review of the environment of human life
was not only a kind of *natural history*—an inventory or systematic *description* of nature—but also an account of origins that claimed to be
nature's true *history* (in the modern sense of that word). However brief
the time-span of Creation was thought to have been, the story did

ascribe to the natural world its own history, divided into a sequence of distinctive periods (the six "days" of the narrative) culminating in the appearance of human beings. It should be clear enough that this conception of world history was—despite the obvious huge contrast in the kind of timescale envisaged—closely *analogous* to the modern view of the Earth's deep history, with its similar succession of major events and new forms of life. To point this out is not to claim that the Genesis account anticipated the truth of the scientific account, but just that the way it was interpreted in the 17th century was *structurally* similar to modern ideas about the Earth's history. The Genesis narrative therefore *pre-adapted* European culture to find it easy and congenial to think about the Earth and its life in a similarly *historical* way.

## NOAH'S FLOOD AS HISTORY

Noah's Flood or Deluge (described later in the book of Genesis) was treated even more clearly as a real historical event: on the chronologists' calculations, it could be dated to more than a millennium and a half after the start of the human drama. Unlike the Creation story, its details did not depend on direct divine revelation. They could have reached Moses—who was believed to be the author of Genesis—through an unbroken line of records or memories stretching back to Noah and his family, who had been on board the Ark and had witnessed the Flood at first hand. So the story of the Flood was subjected to detailed analysis by scholars, who tried to work out what exactly had happened and how. They tried to reconstruct the "*antediluvial*" ("before the Deluge") human world that the Flood had destroyed; how Noah's family had survived the catastrophe in his Ark; and how the "*postdiluvial*" ("after the Deluge") human world had recovered from it. They also conjectured how it might have been caused and how it had affected the Earth itself and its animal inhabitants and other nonhuman features. All this was based on the biblical text, mainly because Genesis was believed to contain the sole authentic historical record of the event (similar non-biblical stories, such as Deucalion's flood in the Greek records, were generally thought to be second-hand accounts derived from the earlier biblical one, or else accounts of later and more local events).

Among the many 17th-century historians who analyzed and commented on the Flood story in this way, the German Jesuit scholar Athanasius Kircher is a good example (just as Ussher has been taken here as a representative chronologist). Kircher was a highly erudite scholar who published on a wide range of topics of interest to his educated readers around Europe; like Ussher he wrote in Latin, making his work accessible to them all. His massive illustrated book on the *The Subterranean World* (*Mundus Subterraneus*, 1668), based on a wide knowledge of the natural sciences of his time, described the physical Earth as a complex system, dynamic but not in any sense a product of history. For example, he speculated how its visible surface features such as volcanoes might be related to its unseen internal structure (he had traveled in Italy and had first-hand knowledge of Vesuvius and

**FIG. 1.5** Kircher's view of the Flood as it subsided, leaving the Ark stranded on the summit of Ararat (right). A matching engraving showed the Flood at its earlier and greatest height, with the Ark floating above the Caucasus (left center), the highest mountain range Kircher knew. These reconstructions combined his interpretation of the textual evidence of the story in Genesis with the natural evidence of what he knew about the Earth's physical geography. (He was well aware that the Ark was not drawn here to scale: like many modern scientific illustrations, this was a *diagram*, albeit in Baroque style.)

Etna). But he did so in rather the same way that the surgeons and physicians among his contemporaries were working out how the visible features of the human body were related to the unseen organs within. Kircher described the Earth's anatomy and physiology, as it were, but he did not describe it as having had any significant major changes or *history* since its initial creation.

The Flood, however, was the great exception. In another massive

**FIG. 1.6** Kircher's "Conjectural Geography" of the world before and after the Flood. This half of his world map marks, conjecturally, "antediluvial" land areas that in the "postdiluvial" world are submerged (*olim Terra modo Oceanus*)—among them the lost land of Atlantis, here placed west of Spain—and, conversely, areas that were formerly under the sea but are now dry land (*olim Oceanus modo Terra*). This illustrated Kircher's claim that the Earth's geography had changed significantly as a result of the Flood. The Earth therefore had a true physical *history*, at least in this respect. His map (in Mercator projection), based on those in contemporary atlases, shows the still barely known Australia as a part of a much larger Antarctica or "unknown southern land" (*Terra australis incognita*), and a similar unknown landmass in the Arctic.

volume, *Noah's Ark* (*Arca Noë*, 1675), Kircher analyzed the Flood historically, using his impressive multilingual skills to exploit all the known ancient versions of the biblical text. He worked out how Noah had built the Ark and embarked with its cargo of varied livestock; how the rising Flood had floated the Ark away and eventually dumped it on the summit of Ararat as the waters subsided; and how the human world had started up again in the postdiluvial period. From the data given in Genesis, he reconstructed, and illustrated in detail, the likely form and size of the Ark. He tried to work out how it could have accommodated even a single pair of every known animal. This gave him a reason to embellish his account with pictures of a wide range of these living animals (in effect, giving his readers a "natural history"). Since the Flood was said to have been worldwide, he also calculated how much extra water would have been needed to raise sea level globally, enough to cover the tops of the highest known mountains; and he speculated about where it might have come from and gone to, unless, implausibly, it was created and then eliminated specially for the occasion.

What is most significant in the present context is that Kircher, like some other scholars, also conjectured that the distribution of land and sea before the Flood might have differed from the form of the continents and oceans after that great event. The Flood might have changed the physical Earth substantially, in no less real a sense than it had changed the human world. At least at this point he was claiming, in effect, that the Earth had a true physical *history* to parallel its human history. However, as the title of his book implied, his erudite analysis was focused primarily on Noah and his Ark, and only secondarily on the physical effects of the Flood itself. His work belonged, in the main, to the same world of thought as that of scholarly chronologists such as Ussher: history was primarily a *human* story, and by modern standards a brief one.

## THE FINITE COSMOS

One practical advantage of the artificial Julian timescale, as chronologists saw it, was that it spanned a total period long enough to accommodate any plausible calculation of the date of Creation at one end, and any anticipated date of the ultimate completion of world history at the other end, leaving plenty of virtual time to spare, as it were, at both ends. It was this that made it convenient as a dimension on which rival chronologies could be plotted and compared. But it also highlights what is, to modern eyes, surely the most unfamiliar feature of Ussher's (and Scaliger's) kind of chronology. This was not that it was very short by our modern scientific standards (though extremely long in human terms), but that it outlined a world history that had *finite limits, both past and future*. In this it was strikingly similar to the "closed world" of the traditional *spatial* picture of the cosmos—with the Earth at its center and all the stars around its periphery—which had been equally taken for granted until astronomers such as Copernicus, Kepler, and Galileo began to open it out into a spatially infinite universe. However, Kircher and many other scholars in his time remained sceptical about that new picture of the cosmos, which they felt had yet to prove itself.

Ussher and most of his contemporaries believed that they were living in the world's seventh and last age. Its final End was widely thought to be imminent, or at least it was expected in the foreseeable future. One common opinion was that the world might end with the completion of exactly six millennia from the Creation (that is, on Ussher's figures, in 1996!). This matched Ussher's calculation that the pivotal point of the Incarnation had been precisely four millennia after the Creation (it had long been recognized that the traditional scale was not quite correct on the real date of Christ's birth, which was put at 4 BC). Such precision, heavily laden with symbolic meaning, made Ussher's figure of 4004 BC particularly attractive to many of his contemporaries; he was not the first or the only chronologist to propose it.

Ussher emphasized and was proud of his achievement, yet he was well aware that his claim was controversial. As already mentioned, many different dates for the Creation were proposed, but not all chronologists were convinced that *any* such date could be fixed. Ever since the early centuries of the Christian (or Common) era, some scholars had pointed out that the Sun, the apparent movement of which defines ordinary days, had not been created until the fourth "day" of the Genesis story. So it had often been suggested that the seven "days" of Creation might not denote periods of twenty-four hours at all. Instead they might represent divinely significant moments, rather like the future "day of the Lord" in the recorded sayings of the Jewish prophets (our use of phrases such as "in Darwin's day" is indefinite in rather the same way). If so, the "week" of Creation might have been of indeterminate duration, and its starting and ending dates might be even more uncertain. In other words, this biblical text, like others, was seen to require interpretation. Its meaning could not simply be read off unambiguously, as if the plain or "literal" meaning was self-evident and beyond argument. This recognition that scholarly judgment was needed, to interpret the meaning of texts, led chronologists and other historians to develop methods of *textual criticism* that continue to underlie historical (including biblical) research to the present day; "criticism" was of course used here in the same sense as in artistic, musical, or literary criticism, without any necessarily negative connotations.

The interpretations of 17th-century scholars may strike us now as extremely literal in character, but this is partly because they were taking the biblical texts seriously as *historical* documents. However, the strongly marked *"literalism"* of their approach to the Bible, far from being an ancient tradition, was a quite recent innovation. In earlier centuries many other layers of meaning—which might be termed symbolic, metaphorical, allegorical, poetic, and so on—had been prominent and generally more highly valued than the literal. But they had sometimes been elaborated so fancifully that, particularly in the Protestant world in the wake of the Reformation, they were downplayed or stripped away altogether, leaving the supposedly simpler "literal" meaning supreme. Yet Protestant scholars, no less than Catholics, conceded and indeed emphasized that their interpretations of biblical texts were intended primarily to elucidate practical *meaning* based on theological understanding, not to impart knowledge of nature.

In the case of the Creation story, for example, what was thought to be of ultimate importance was not its exact date, or the duration of its "days." Far more significant for human lives was its assurance, in effect, that all things had been freely created by the one and only God, who had pronounced them all to be intrinsically "good"; that the sequence of creative actions had not been arbitrary but underlain by the consistent purposes of a caring God; and that no created *thing*—not even angels or other heavenly powers, let alone the Sun or Moon or other natural entities—should be treated as ultimate in value or deserving of worship. Themes such as these had been the stuff of both popular sermons and scholarly commentaries on Genesis, ever since the early Christian centuries. The theological *meaning* of the texts, and their application in the practice of Christian faith, had been emphasized endlessly, taking priority over any use they might have as sources of factual knowledge about the world's origins. (The historically recent rise of literalism, and the continuing primacy of theological meaning in biblical interpretation, are often overlooked or ignored by modern fundamentalists, religious and atheistic alike.)

The exact date of Creation was not the only unresolved problem lurking behind the chronologists' confident dating of world history. Although Egyptian hieroglyphic inscriptions could not be deciphered, there were ancient Greek reports of what had been known at that time. According to these, Egypt's early dynasties stretched back many centuries before the generally favored dates for the Creation. The two sources of alleged evidence—Egyptian and biblical—could not both be correct, so chronologists had to choose between them. Once again, scholarly judgment was needed. It is hardly surprising that the biblical record was treated as the more reliable. The Egyptian record of allegedly pre-Creation history was generally dismissed as political spin, as a fiction devised long ago to bolster the legitimacy or prestige of Egypt's ancient rulers. Equally unsettling, however, were some of the early Chinese records, when research by Jesuit scholars living in China first made them known to other Europeans. These records too suggested a much longer ancient human history than the chronologists' calculations allowed. And ancient Greek reports of Babylonian records, though generally dismissed as fictitious, claimed even greater antiquity for human civilizations.

Most unsettling of all, perhaps, were the conjectures contained in a small book published just after Ussher's huge *Annals*. The author of the anonymous *Men Before Adam* (*Prae-Adamitae*, 1655), which soon became widely known and indeed notorious, used a subtle interpretation of a specific New Testament text to argue that the biblical story of Adam was originally intended to refer to the first Jew, not the first human being. This put a big question mark against all chronologies based on Adam as the starting point for human history. The conjecture had the advantage that it could explain how the human races around the world might have had time to become so widespread and diverse. The sheer variety of humanity had not been fully appreciated by Europeans until, little more than a century earlier, their great exploratory voyages had first taken them around Africa to Asia and across the Atlantic to the Americas. Conversely, however, the conjecture had the disadvantage that it seemed to deny the unity of human-kind in the Christian drama of salvation. For example, it appeared to exclude the indigenous peoples of the Americas from that drama, and thereby denied them fully human status. The claim that there had been "Pre-Adamite" human beings got its author—whose identity had been disclosed as the French scholar Isaac La Peyrère—into trouble with the Catholic authorities, though after renouncing such specula-tions, at least nominally, he lived to a peaceful old age.

## THE THREAT OF ETERNALISM

In the present context, however, the importance of Pre-Adamite ideas is that they added to the impact of the allegedly ancient Egyptian, Chinese, and Babylonian records. They all implied that the totality of human history might be far longer than any conventional Western chronology allowed, stretching back not five or six millennia but per-haps more than ten, or even—if the Babylonian records were to be believed—many tens of millennia. All this was disturbing to conven-tional thinking: not primarily because it put the dating of Creation or the authority of the Bible in doubt, but far more because it seemed to open the door to a much more radical kind of speculation. It sug-gested that ancient Greek philosophers such as Aristotle and Plato, whose ideas on other topics had long been revered in Europe, might have been right in this: they were taken to have claimed that the uni-verse, and with it the Earth and human life, are not just extremely an-cient but literally *eternal*, without any created beginning or final end. This was profoundly disturbing, because to deny that human beings are in some sense *created*, and therefore morally answerable to their transcendent Creator, seemed equivalent to denying that they have any ultimate responsibility for their actions and behavior. It seemed to threaten the very foundations of morality and society.

At first glance, this *"eternalism"* (as it has since been named) might seem to anticipate the modern picture of a history of the Earth and the cosmos measured in billions of years, in sharp contrast to the chro-nologists' short and finite story measured in mere thousands. But the apparent modernity of eternalism is deceptive and deeply misleading. In fact, a "young Earth" and an eternal one, which were the only two alternatives considered in the 17th century, were *equally un-modern*.

Both assumed that human beings have always been and will always be essential to the universe. Although the chronologists' short and finite history of the Earth (and of the cosmos) included a very brief pre-human setting of the scene, it was otherwise wholly a human drama from start to finish. But the eternalists' picture, likewise, was of an Earth (and a cosmos) that had never in the past been without human beings, or at least some rational Pre-Adamites, and never would be in the future. Those who argued for the authenticity of extremely early human records from Egypt, China, or Babylon, way back beyond the usual range of plausible dates for Creation, assumed that even these were just the most ancient that had happened to survive. They took it for granted that there must have been a long or even infinite sequence of still earlier human cultures, of which all traces had been lost in the mists of time.

So the *infinitely* ancient Earth (and universe) of eternalism did not anticipate the modern scientific picture of an immensely lengthy but *finite* history of the Earth (and of the universe). Yet at the time, in the 17th century and even later, eternalism did offer a radical alternative to what was then the culturally dominant picture of a probably brief and certainly finite universe. Eternalism was widely regarded as subversive, socially and politically as well as religiously. So it generally remained, as it were, underground: it was most often visible when it was attacked by its orthodox critics, rather than being expressed openly by its un-orthodox proponents. What was perceived as the radical threat to human society posed by eternalism goes a long way to account for the dogged defense, in some circles though not all, of the "young Earth" derived from a very literal interpretation of the Creation story in Genesis. Conversely, however, eternalists often had their own—religiously sceptical or even atheistic—agenda to promote. So this was certainly not a straightforward struggle of enlightened Reason against religious Dogma. There were strong "ideological" issues at stake on *both* sides of the argument.

On a global scale, however, the idea of an indefinite or even endless sequence of human lives, as implied by eternalism, had been the norm rather than the exception. Most pre-modern societies around the world embodied in their cultures an assumption that time—or rather, the *history* that unfolds in time—is repetitive or in some sense cyclic, not arrow-like or uniquely and irreversibly directional. Underlying this assumption, and making it seem common sense, was the universal experience of the cycle of individual lives from birth through maturity to death, repeated from one generation to the next. This was powerfully reinforced by the annual cycle of the seasons, which in most premodern societies was an extremely powerful determinant of human lives. Together they fostered a similarly cyclic or "*steady-state*" view of human cultures, of the Earth, and of the universe as a whole. Against this background, the idea that the world has had a unique starting point and a linear and irreversibly directional *history*—an idea that first emerged in Judaism and was extended in Christianity (and later in Islam too)—stands out as a striking anomaly. Each of the Abrahamic faiths condensed its directional view of history into an annual cycle of fasts and festivals (Passover, Easter, etc.), which replicated the cosmic picture in miniature on the more accessible scale of ordinary human lives. But the larger-scale picture remained paramount, namely that humanity, the Earth, and the universe have jointly had a true *history*, with an irreversible arrow-like direction to it.

This strong sense of history gave the Judeo-Christian tradition an underlying structure that is closely analogous to the modern view of the Earth's deep history (and cosmic history) as similarly *finite and directional*. More specifically, the science of scholarly chronology, as a way of plotting human history with quantitative accuracy and of dividing it into a qualitatively significant sequence of eras and periods, was closely analogous to the modern science of "*geochronology*," which tries to give a similar kind of precision and structure to the Earth's deep history, dividing it in the same way into eras and periods. Whether these are "mere" analogies, or something much more, is a question that the rest of this book will explore.

To summarize: the history of the universe, the Earth, and human life itself was traditionally conceived in the West as having been very brief in comparison with the modern picture. But this is a relatively trivial difference: the quantitative contrast is less significant than the qualitative similarity. What is not trivial is that the scholarly history represented by chronologists such as Ussher was almost exclusively based on *textual* evidence (the astronomical evidence of past eclipses, comets, etc. also came from textual records). Even in the historical analysis of Noah's Flood by scholars such as Kircher the textual evidence was dominant and the use of natural evidence was marginal.

At much the same time, however, and still in the 17th century, other scholars were beginning to bring the natural evidence much more substantially into debates about the Earth's own history, yet without seeing any obvious need to extend the timescale on which it had played out. This is the subject of the next chapter.

**2**

# Nature's Own Antiquities

## HISTORIANS AND ANTIQUARIES

In retrospect it might seem obvious that evidence from the natural world—such as rocks and fossils, mountains and volcanoes—ought to have undermined the idea of a "young Earth" from the start. But in fact the significance of such features was far from obvious, and for some very good reasons. Among these was the sheer novelty of the idea that nature might have had any real history, after its chief components had been put on stage during the "week" (literal or not) of Creation. Apart from the much later and unique mega-event of the Flood, the natural world was taken to have been a stable backdrop throughout the ongoing drama of human history. That nature might have had its own dramatic action began to seem plausible only when the ideas and methods of historians were *transposed* into the natural world, from culture into nature. History—human history—was a flourishing field of scholarship in the 17th century, and its variety and high standards provided fertile ground for this crucial transfer.

James Ussher's 4004 BC was not the only date calculated by chronologists for the week of primal Creation. But nor were chronologists the only people doing historical work in the 17th century. Chronology was just one rather specialized kind of history. It was multilingual and multicultural in its sources; it interpreted world history in terms of the Christian narrative of cumulative divine self-disclosure or "revelation"; and generally it set out its results in the form of "annals," or chronicles of events arranged in a year-by-year sequence with as much precision as chronologists could muster. Other scholars wrote histories of other kinds, often more secular in character and taking ancient Greek and Latin writers as their models. These were histories of particular places or peoples, or of specific periods or episodes in the past, or of the lives and influence of individuals of outstanding importance. Like the chronologists, other historians often divided the past into periods of distinctive character, or they adopted periods already in general use. Periods could be useful for descriptive purposes, even if their dates were not defined precisely. For example, the "Middle" ages or mediaeval period filled the centuries between the decline of the "Ancient" or Classical world of Greece and Rome and the Renaissance or rebirth that marked the start of the "Modern" world.

Documents and books, stored in archives and libraries, were essential in almost all kinds of historical work. Like the chronologists, other historians adopted rising standards of scholarly rigor and an increasingly critical evaluation of the reliability (or otherwise) of their sources. Secular textual records, no less than scriptural ones, required critical scrutiny. Records contemporary with the events themselves were most highly valued, and historians learned how to detect telltale signs of anachronism: the documents might be later forgeries, which in turn might have significant political implications. Conversely, many historians believed that valuable clues about the earliest periods of human history, back before the relatively well-documented ages of Greece and Rome, might be preserved in the apparently unpromising form of myths, legends, and fables. Stories of gods and superhuman heroes might in reality be garbled accounts of great ancient rulers and exceptional natural events. These stories might seem at first sight incoherent or implausible, but with suitable demythologizing—a method termed "*euhemerist*" after its much earlier exponent the ancient Greek author Euhemerus—they might throw light on the very earliest "fabulous" or "mythical" stages of human history.

However, historians also made increasing use of other kinds of evidence. Textual documents could be supplemented with other records of what had happened in the past. For the Classical past of Greece and Rome, for example, there were the inscriptions found on ancient buildings, or dug up in their ruins: no less textual than conventional documents, but often supplementing them by providing important new information about distant past events. Then there were the coins found on ancient sites, which often helped with dating because they preserved bits of text combined with portraits of ancient rulers and other significant images. Other artifacts, even without any text attached, provided further evidence about great events or ordinary everyday life in these much admired ancient cultures. They ranged, for

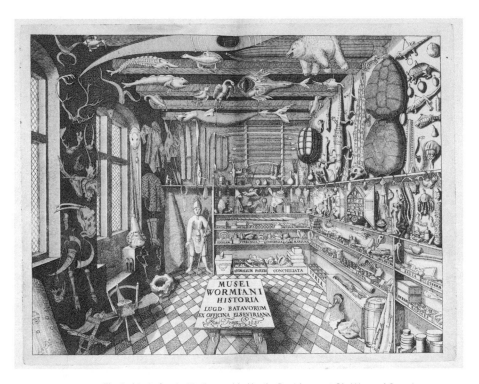

FIG. 2.1 The "cabinet of curiosities" assembled by the Danish savant Ole Worm of Copenhagen. It included a wide variety of interesting or puzzling objects of all kinds, natural and human, all carefully classified. Most of his fossils, for example, would have been stored on the lower shelves, and classed as *Lapides* ["Stones"]. This engraving (here greatly reduced in size) formed the spectacular frontispiece or visual summary of the book describing and illustrating Worm's collection (*Museum Wormianum*, 1655); being in Latin, it was accessible to educated people throughout Europe.

example, from Greek vases and Roman sculptures to *"monuments"* such as Greek temples and Roman theaters. All such artifacts, which together supplemented documentary sources, were known as *"antiquities."* The smaller and more collectable kinds (in modern terms, just "antiques") often featured prominently—alongside an amazing variety of other rare, curious, or puzzling objects, both natural and human—in the "cabinets of curiosities" or private museums assembled by scholars, particularly those who called themselves antiquarians or *"antiquaries."*

There was no reason why antiquities could not be used as historical evidence, even in the total absence of textual sources. For most of Europe, evidence for the times before the Romans arrived with their literate culture was confined to undateable artifacts such as stone tools or weapons found lying on the ground, or bronze objects and pottery dug up from ancient tombs, or striking but enigmatic monuments such as the huge stone circle of Stonehenge in the south of England. It seemed possible that some of these artifacts might date from the same centuries as early literate cultures elsewhere, such as the Classical Greek world around the Mediterranean. But it was conceivable that some other artifacts were older than any textual evidence anywhere (apart from the very sparse early biblical records, and the early myths preserved by other cultures, all of which were controversial). So the artifacts studied by antiquaries, even if undateable, might help to throw light on the earliest periods of human history, back before the times from which *any* reliable textual documents survived. In effect, they might replace, and not merely supplement, the more traditional sources of historical evidence.

## NATURAL ANTIQUITIES

There was also no reason why these antiquities could not be supplemented in turn, or even replaced, by other ancient material objects, which were not artifacts because they were not human but natural in origin. Nature might, metaphorically, have its own antiquities. The most striking objects of this kind were the sea shells that in some regions could be picked up on the ground, far from the sea and sometimes high above it. These *"natural antiquities"* had already been noticed and commented on, back in Classical times. In the 17th century

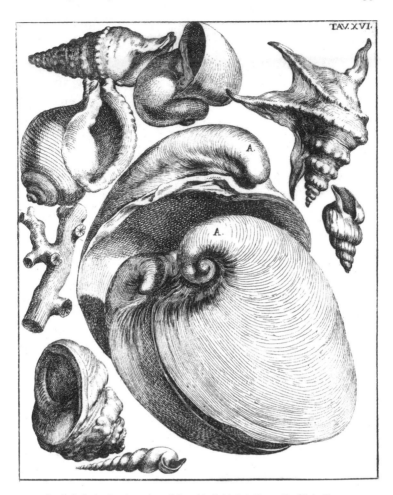

**FIG. 2.2** Fossil shells (and a piece of coral) found in Calabria in the south of Italy. These were among the many engravings published by Agostino Scilla in *La Vana Speculazione* (1670). They supported his claim that such objects were the remains of shellfish and other creatures that had once been truly alive. They were so similar to the shells of molluscs, sea-urchins, corals, etc. in the nearby Mediterranean that any other explanation of them was, he claimed, no better than a "vain speculation." (The shells are fitted economically into the space available on the expensive copper plate from which the engraving was printed.)

many scholars, like their forerunners in the ancient world, believed these shells showed that in the distant past the seas had extended far beyond their present limits. For example, the Sicilian scholar (and professional painter) Agostino Scilla published an account of those he had collected on his native island and in adjacent parts of Italy. He claimed that his first-hand observations showed clearly that they were the shells of shellfish that had once been truly alive. He brusquely dismissed any alternative as a "vain speculation" that flew in the face of "sense."

The most obvious point in known human history to which such a major change in geography could be attributed was Noah's Flood. This was the only episode of its kind for which, in the opinion of Kircher and many other scholars, there was trustworthy first-hand testimony recorded in a reliable documentary source. If, as the narrative in Genesis appeared to record, the Flood had been worldwide in extent, it could perhaps account for these sea shells found so widely, far from the sea and often high above it. The Flood or Deluge provided a "*diluvial*" explanation of these otherwise puzzling natural objects. Those who interpreted them in this way were not forced to do so by any narrow-minded biblical literalism, let alone by any repressive church authority. The explanation seemed, at least at first, just as natural as the sea shells themselves. And since it was a *historical* explanation, uncertainty about how the Flood might have been caused was no reason to reject it. The historical reality (or otherwise) of the Flood was recognized as a separate issue from the discovery of its cause (which was often thought to have been natural, although at the same time its ultimate purpose was said to have been divine).

Those scholars who attributed these sea shells to Noah's Flood often hoped that this would reinforce the trustworthiness of Genesis, and indeed of the Bible as a whole. But those who opposed such objectives, and doubted or denied that the changes had been due to the Flood, could still agree that the shells were reliable natural evidence for major changes of geography. The changes might have been so far back in human history that they were recorded, if at all, only in the garbled form of myths and legends. But these could be demythologized in euhemerist fashion. As Kircher and others claimed, large parts of the present continents might once have been under the sea, just as, conversely, Plato had recorded the legend that the sea now covered the formerly inhabited land of Atlantis (the possible location of which was much debated). Once again, how these geographical changes might have been caused was a separate issue. In any case, natural antiquities such as sea shells found far inland were assiduously collected by many scholars, and stored in their cabinets alongside the more conventional antiquities that were unquestionably human artifacts. All alike were potential evidence for early human history.

Not all alleged traces of the Flood or other major geographical changes were as easy to interpret as Scilla's sea shells. These were just one category within a much larger and more diverse group of objects known collectively as "*fossils*." The word meant simply "things dug up," and it included *all* kinds of distinctive objects or materials found on, or more usually below, the surface of the ground (this original meaning of the word survives in our term "fossil fuels" for the coal dug up, and the oil pumped up, from the depths of the Earth). The "fossils" collected by 17th-century scholars, and conserved in their cabinets or museums, included a wide variety of such objects. They ranged from quartz crystals and other minerals at one end of a spectrum to obvious sea shells at the other end. In between were a lot of more or less puzzling objects with more or less resemblance to living plants and animals (or to bits of them). So the problem was not to decide *whether* "fossils" were organic in origin or not, but to decide *which* were the remains of organisms (or parts of them) and which were not. The question was, in which "fossils" was the resemblance to plants or animals due to their origin as parts of such living beings, and in which others was any resemblance accidental or a matter of chance? Only those that were truly organic in origin could be regarded as nature's own antiquities, and therefore be used to supplement, or even replace, other forms of evidence about the *history* of humanity and its terrestrial environment.

In fact, the problem was not as straightforward as this suggests. The "more-or-less" resemblances between many "fossils" and living plants and animals were widely attributed neither to chance nor to a simple causal connection, but to a fundamental analogy between the inorganic and organic realms of nature. The inorganic or mineral world was widely believed to generate forms that, although they had never been truly alive, bore some resemblance to, or had some "correspondence" with, the forms generated by the organic or living world: for example, there were the vaguely fern-like mineral forms (in modern terms *dendritic* markings) that were often found on the surfaces of rock when blocks were split open. The conception of nature that justified this kind of explanation is now quite difficult to understand, but it was widespread and even dominant in the 17th century. In the specific case of "fossils," its strength as an explanation was that it seemed to account naturally for all the puzzling features of these puzzling objects.

In brief, their *form* often differed from that of any known living animal or plant; their *substance* was usually that of minerals rather than organic materials; and their *position*, "dug up" from beneath the surface, suggested that they might have grown underground like minerals, and not within an organism living in some now vanished sea.

This was the kind of "vain speculation" that Scilla hoped to dispel, deploying "sense" or observation to replace it with the explanation that fossil shells were the remains of shellfish that had once been truly alive. But Scilla had a relatively easy case to argue: his "fossils" were at the easy end of the spectrum. They were similar in form to shellfish living in the Mediterranean; they were little different in substance from shells lying on the seashore; and they were found close to the present sea. Most other "fossils" were much more difficult to understand. Many were unlike any animals or plants known alive, at least in detail; they were generally "petrified" or stony in substance; and they were often found enclosed in hard rocks far inland and high above sea level. For them, an explanation in terms of a subtle "correspondence" between the living and non-living worlds could seem much more plausible and not at all a "vain speculation." "Fossils" were therefore a focus of lively debate all across Europe, because fundamentally different conceptions of nature were at stake. In the late 17th and early 18th centuries, arguments about the interpretation of the various kinds of "fossils" were almost as intense as those concerned with, say, the basic forces of nature, the ultimate structure of matter, or the essential character of life itself.

Such debates were not confined to scholars working mainly with books and other textual materials, or antiquaries studying ancient artifacts. Fossils, like animals and plants, were regarded as objects of natural history, and were therefore studied by those known as "*naturalists*" (a term that had none of its modern overtones of amateurism). Questions involving natural causes, such as the origin of fossils or of the waters of the Flood, were the province of "philosophers," and more particularly those who called themselves "*natural philosophers*." These categories were not sharply distinguished, because all these people saw themselves as contributing to the interconnected bodies of disciplined knowledge that were known as "sciences" (in the broad and plural sense that, as already mentioned, has remained in use to the present day except in the English-speaking world). They all regarded themselves, and were regarded by the wider public, as "*savants*" (knowledgeable people), an umbrella term that was widely used into the 19th century (using it here, where appropriate, will avoid the anachronistic use of the much narrower English term "*scientist*," which was not coined until well into the 19th century and not used generally until well into the 20th).

The opportunities for discussions among savants of all kinds were greatly facilitated by the scientific bodies that were founded in the 17th century in several European cities, notably in the capitals of two of the political superpowers, France and England, and by the publication of the first regular newsletters or periodicals specifically written for such people. These new places of debate were the Académie des Sciences in Paris, where the *Journal des Savants* was also produced, and the Royal Society and its own *Philosophical Transactions* in London ("philosophical" here meant "natural-philosophical" and was roughly equivalent, in modern terms, to "scientific"). But much scientific debate also continued in more traditional ways, when savants met in the course of their travels around Europe, at other times through the letters they wrote to each other, and in what they published and distributed in their books and pamphlets.

## NEW IDEAS ABOUT FOSSILS

Two such savants, whose studies of the fossil problem became especially significant, were the Danish physician Nils Stensen (more usually known as Steno, the Latinized name under which he published his work) and the Englishman Robert Hooke. Steno, having studied in Denmark, Holland, and France, was appointed to an important medical position in Florence, the capital city of the powerful state of Tuscany in central Italy (such cosmopolitan careers were common at this period, just as they are among scientists in the modern world). Here he joined a group of savants who had been inspired by the great Galileo earlier in the century. Hooke, somewhat in parallel, worked in London at the new Royal Society, which was modeled in part on the group in Florence; he was employed to instruct and entertain its members with experiments and demonstrations. The members of these bodies, and many other savants all around Europe, were exploring new ways of studying the natural world, often interpreting it in material and mechanical terms inspired by the burgeoning world of technology. This was the common intellectual environment in which Steno and Hooke both became involved in the ongoing debate about "fossils," so it is not surprising that they reached similar conclusions (charges of plagiarism later flew in both directions, but are not supported by the historical evidence).

In 1667 Steno published a short report on his dissection of the head of a large shark that happened to have been landed on the coast of Tuscany. He included in it what he called a "digression" on the well-known fossils called *glossopetrae* ("tongue-stones"). These objects were somewhat tongue-shaped but closely similar to sharks' teeth, though much larger in size. But they were petrified or stony and they were found on land, embedded in solid rock. Steno thought them so significant that he planned a major work dealing with the interpretation of "fossils" in general. He only published a brief trailer (*Prodromus*, 1669) before he was recalled to medical work in his native Copenhagen; he later returned to Italy, but having become a Roman Catholic and been ordained as a priest he had other duties and priorities, and he published nothing more on this topic (that he abandoned it because he was scared about its implications for his religious faith is a myth concocted by modern commentators hostile to his or any religion). But what he had published became well known throughout Europe and generated lively debate among savants. In London it was recognized as setting out conclusions similar to those already reached by Hooke. In his book *Micrographia* (1665) Hooke had described the astonishing new world of small-scale nature revealed by the recently invented microscope. Among many spectacular illustrations of minute objects—a tiny flea, the compound eye of a fly, and so on—he had also depicted the microscopic appearance of petrified fossil wood and of a piece of charcoal, showing that they had a similar microstructure of little "cells." This was Hooke's equivalent of Steno's sharks' teeth: "fossils" that were, they each claimed, certainly organic in origin. At least *these* objects, they argued, were genuine natural antiquities that could properly be used, along with sea shells found far from the sea, as clues to nature's own history.

Both savants had first to give reasons for rejecting the idea of intrinsic "correspondences" between the organic and mineral realms. They appealed to the traditional principle that "nature does nothing

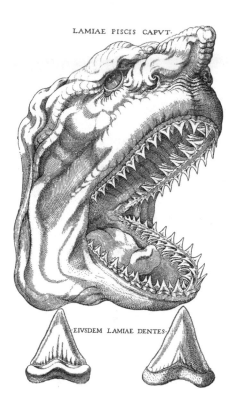

LAMIAE PISCIS CAPVT.

EIVSDEM LAMIAE DENTES.

**FIG. 2.3** Steno's illustration (1667) of the head of a shark, showing the very large number of teeth, most of them waiting as it were to be brought into use. Below are internal and external views of one of the teeth. In a book published soon afterwards, Steno compared these teeth with the well-known "fossil objects" called *glossopetrae* or tongue-stones, as an example of how all such "fossils" should be interpreted.

in vain": objects that obviously made possible the lives of sharks, shellfish, and trees would not be formed by nature just in order to exist forever embedded in a rock. A closely related principle was drawn from "*natural theology*" (the branch of theology that analyzes claims about the relation between God and the natural world, including human nature, complementing the "revealed theology" that evaluates claims about God's self-disclosure in human history). This principle was that all the forms of animals and plants had been divinely designed to enable them to follow appropriate ways of life, which they certainly could not do inside a rock. Steno also analyzed the difference between the growth of a tooth in a shark's jaw and the growth of a crystal underground in a rock: there was no true analogy between the two kinds of growth.

Both savants then had to account for the differences in form, substance, and position between their "fossils" and any living animals or plants. They and their contemporaries were developing theories

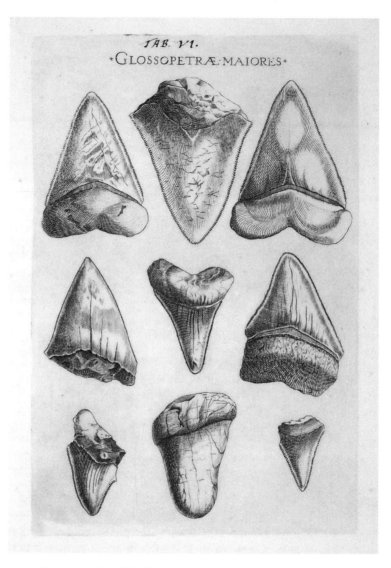

**FIG. 2.4** Steno's illustration (1667) of large *glossopetrae* ('tongue-stones') found in solid rock. In his *Prodromus* (1669) he argued that these "fossils" were truly the teeth of sharks much larger than any known alive, dating from some very early period of history: they were *fossils* in the narrower modern sense of the word.

about matter in general that made questions about *substance* relatively straightforward. They could readily imagine how wood or sharks' teeth or shells could have been turned into stone by the infiltration of tiny particles of mineral matter, percolating in solution through the rock and being precipitated within the original organic matter or replacing it altogether. And the solid rocks enclosing the fossil objects could have been produced in much the same way, consolidated from the soft sediments that Steno claimed they must have been originally. Hooke, who went on to consider a much wider range of "fossils" than Steno, also worked out why, conversely, in some such objects there is no shell material at all. After the sediment became a solid rock, percolating water could have dissolved the original shell away, leaving nothing but an empty "mold" like those made by jewelers for casting forms in gold or silver. Yet even a mold would still preserve the shape of a genuine shell produced by a real shellfish.

The *position* of sharks' teeth and sea shells on land, often far from the sea and high above it, was less easy to explain. Steno conjectured that the layers or "*strata*" of rock had been deposited as soft muddy sediment at a time when the sea was far above its present level—he thought this would have been during the biblical Flood—and subsequently left high and dry when the waters receded. Hooke, on the other hand, invoked past earthquakes, which might have heaved parts of the Earth's crust up from the sea floor to form new land areas. Both conjectures, however, raised further problems. Hooke, like many other commentators, dismissed the Flood as having been too brief an interlude to produce the observed effects. But the earthquakes he invoked were criticized by other naturalists, since his native land was replete with the relevant fossils but usually free from earthquakes (a few English earthquakes at just this time attracted much attention precisely because they were so unusual).

The problems raised by contrasts in the *form* of fossils and their apparent living counterparts were more acute for Hooke than for Steno. The fossil *glossopetrae* were closely similar to the teeth of sharks; the best-known examples were much larger, but the discrepancy was lessened by the huge shark that had prompted Steno's research in the first place. And the fossil shells that he (and Scilla) found in the Italian rocks were also quite similar to living shellfish. Hooke, on the other hand, analyzed a much wider range of English "fossils." For example, he had

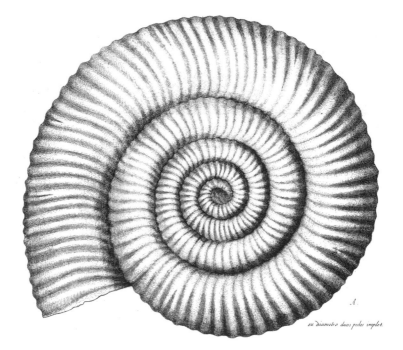

*ex diametro duos pedes implet.*

**FIG. 2.5** A giant ammonite (two feet [60 cm] across), as depicted in Martin Lister's massive illustrated *[Natural] History of Shells* (*Historia Conchyliorum*, 1685–92). Lister, a physician and early member of the Royal Society in London, doubted whether this and similar "fossil" shells were truly the remains of animals that had once been alive, because they were so different in form from any living shellfish—about which he knew more than almost anyone else at this time—and because they seemed to be composed only of rock, without any shell material (in modern terms, they were "molds"). The shells nearest in form to ammonites were those of the "pearly nautilus" from the tropical seas around the East Indies (now Indonesia).

to contend with the varied and beautiful "*ammonites*" much prized by collectors of curiosities, which were unlike the shells of any known living shellfish. But he realized how little was known about the animals and plants living in remote parts of the world—every long-distance voyage or expedition brought many new and unknown forms back to Europe—so he thought it reasonable to predict that those known only as fossils might eventually be found alive. Alternatively, he conjectured that some species might have changed in form in the course of time, just as new breeds of domestic animals had been developed (his ideas on this point have a misleading similarity to much later ideas about evolutionary change). These problems continued to puzzle Hooke as he lectured at the Royal Society on "fossils" and earthquakes—which remained topics of great interest to its members—over the next thirty years. They also puzzled many of his contemporaries.

## NEW IDEAS ABOUT HISTORY

However, nothing about Steno's *glossopetrae*, nor the ammonites and other "fossils" that Hooke discussed, led either of them to question the kind of timescale for the Earth that almost all savants took for granted (and that chronologists tried to quantify precisely). Nor did they doubt that almost all this span of time had been *human* history. For example, Steno pointed out that the great abundance of the tongue-stones famously found on the island of Malta was no argument against either their organic origin or a brief timescale, because even a single living shark produced some 200 teeth (in use or, as it were, in reserve). He also noted that blocks of rock containing fossil shells, quarried locally, had been used by the ancient Etruscans for building the walls of the hill-town of Volterra in Tuscany, even before the Romans conquered that region and wiped out Etruscan culture. This showed that both the rocks and the fossils were not just pre-Roman but pre-Etruscan, which put them far back in ancient history. So Steno argued that their formation must date from still further back, even perhaps as far back as the biblical Flood. Far from having to squeeze his evidence into an uncomfortably brief time, he anticipated that his readers would need to be convinced that these natural antiquities, unlike many ancient human artifacts, could have been preserved in such good shape through so *long* a span of time.

Like Steno, Hooke assumed that all the events he reconstructed had taken place within the span of human history, however remote. He suggested that England might have suffered in the distant past from very powerful earthquakes (as Steno's Italy still did), which could have heaved rocks and their fossils high above sea level. But he anticipated that further evidence for this would be found not in nature but in ancient *human* records, namely in fables and legends that could be demythologized (in the usual euhemerist fashion) as garbled accounts of ancient earthquakes and volcanic eruptions. He insisted, against the objections of some of his contemporaries, that ammonites were formed by shellfish as yet unknown. Since some were of giant size, and they were nearest in form to the beautiful and highly prized shell of the tropical "pearly nautilus," he thought England might once have enjoyed a tropical climate. But here again he anticipated that further evidence for this would be found not in nature but in ancient documentary records. He also suggested that it might be possible to "raise a chronology" from fossils. But by this he simply meant that they could complement the textual sources that chronologists used, or, at most, replace them and extend the record into those earliest periods of human history for which no unambiguous documents survived. He was aware of the Egyptian and Chinese records of a longer history than was allowed for by most chronologists; but even if they were genuine—which he doubted—they merely extended by a few millennia what was still *human* history.

In considering how "fossils" should be interpreted, neither Hooke's ideas nor Steno's were cramped, let alone distorted, by their common assumption that the traditional outline of world history was broadly on the right lines, and its timescale the right order of magnitude. Yet they did each introduce an important new element into the ongoing debate about the history of humanity and its natural environment. Both deliberately transposed the ideas and methods of *historians* into the natural world, without seeing any need to extend significantly the kind of timescale of global history that they and their contemporaries took for granted.

Steno used the rocks and fossils he observed in Tuscany—a region he thought would prove typical of the Earth's surface—to reconstruct a historical *sequence* of natural events. He matched this with the biblical narrative of Creation and Flood without, in his own eyes, any awkwardness at all. In the hills around Volterra he found two distinct sets of rock, one lying above the other, and each with strata that were horizontal in some places but tilted in others. He inferred that they had all been deposited originally in horizontal layers, but that in some places they had later collapsed into a tilted position. The upper set contained fossil shells; the underlying set, which was obviously older, had none. So he inferred that the lower set dated from the time of Creation, before there were any living things, whereas the upper set dated from a later period, probably the time of the Flood. He had no difficulty in concluding that these natural antiquities either confirmed the biblical narrative of the earliest periods of history or were at least compatible with it: as he put it, either "scripture and nature agree" or "nature

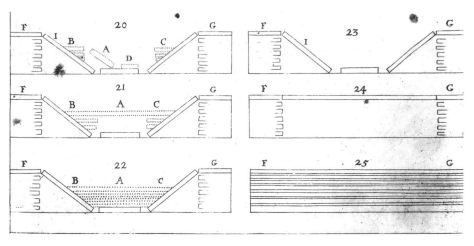

**FIG. 2.6** Steno's diagram to illustrate his reconstruction of the physical history of Tuscany, published in his *Prodromus* (1669). The six "sections" or vertical slices through this part of the Earth's crust could be read in two complementary ways. The numbering denotes the order of investigation, from the observable present state of the rocks (20) back to the inferred original state (25); but the text traces Steno's inferred *history* in real time in the opposite order, from the original (25) forwards to the present (20). The rocks of the older set (strata ruled in continuous lines), without fossils, were deposited in horizontal layers (25) but then undermined (24) until the upper layer collapsed into tilted positions (23). Later, in a similar but distinct sequence of events, the rocks of the younger set (strata ruled in dotted lines), with fossils, were deposited on top of the older set, in horizontal layers (22), which were then undermined (21) until the uppermost collapsed into their present position (20). This diagram was drawn in an abstract geometrical style that recalls Galileo's physical research, a tradition that inspired Steno and his colleagues in Florence.

proves it and scripture does not contradict it." (That the strata *within* each of his two sets of rocks also represent a sequence of events on a smaller scale—each layer having necessarily been deposited before the one above it—was an inference so obvious that it hardly deserves its much later elevation into a formal "principle of superposition," and Steno deserves no special credit for using it as a matter of course.)

Steno's summary of this historical sequence shows how his method of reasoning paralleled that of the chronologists. Just as they pieced the evidence together by starting with the more recent and relatively well-documented past (in Ussher's case, the Roman period) and then penetrating back into more obscure earlier times, so Steno began with the present state of Tuscany and reasoned back from it. Just as the chronologers then reversed direction and set out their reconstructed history forwards in real time, as accurately dated "annals," so Steno

> There is no Coin can so well inform an Antiquary that there has
> been such or such a place subject to such a Prince, as these [fossil
> shells] will certify a Natural Antiquary, that such and such places
> have been under the Water, that there have been such kind of An-
> imals, that there have been such and such preceding Alterations
> and Changes of the superficial Parts of the Earth: And methinks
> Providence does seem to have design'd these permanent shapes,
> as Monuments and Records to instruct succeeding Ages of what
> past in preceding [ages]. And these [are] written in more legible
> Characters than the Hieroglyphicks of the ancient *Egyptians*, and
> on more lasting Monuments than those of their vast Pyramids
> and Obelisks.

FIG. 2.7   A revealing quotation from a lecture that Hooke gave at the Royal Society in London in
1668. The work of a "natural antiquary" studying fossils was, he suggested, closely analogous to
the work of a conventional antiquary studying *human* artifacts. In another lecture he suggested
that fossil shells might date from an earlier time than even the oldest "monuments," but it is
not clear that he thought they were any older than the earliest "mythical" or "fabulous" ages of
human history.

reconstructed the successive phases of the Earth's history in unquantified but equally real time, from the most remote period to the present. Steno's brief essay on this topic in his *Prodromus* was a striking
example—and subsequently an influential one—of how the Earth's
early history could be reconstructed, using rocks and fossils as "natural antiquities" to complement, supplement or even replace the textual
evidence assembled by chronologists.

Hooke, in an equally significant move, deliberately applied the antiquaries' methods to the study of the Earth. The *"natural antiquary"*
could use rocks and fossils as historical evidence of ancient changes in
geography, even before those periods of human history from which
any documentary records survived. Rocks and fossils were *nature's*
monuments and *nature's* coins. These were metaphors, but far more
than "mere" metaphors: they did serious explanatory work. Rocks
and fossils could even be treated as nature's own documents, written as it were by nature's own witnesses to events long past. But then,
like ancient human records, they would have to be deciphered and

their meaning interpreted. "Nature's Grammar," the language of nature, had to be learned before nature's antiquities could be used to reconstruct history; otherwise they remained as uninformative as the hieroglyphic writing of the ancient Egyptians, which was well known but undeciphered.

## FOSSILS AND THE FLOOD

Steno and Hooke were not isolated geniuses; they were engaged in lively debate with their contemporaries. But they did provide other naturalists with fruitful models on which further research could be based during the rest of the 17th century and into the 18th (Hooke's ideas were less influential than Steno's, partly because his lectures were not published until after his death and then only in English). One of their most active younger contemporaries was the English physician John Woodward, who amassed one of the finest collections of fossils anywhere; at his death he bequeathed them to the university in Cambridge and endowed lectures that were to expound his ideas about them. (The fossils and the lecturer became, respectively, what is now a major geological museum and a named professorship heading a distinguished department; I was trained there as a palaeontologist by a 20th-century "Woodwardian professor.") Woodward's major book, *An Essay on the Natural History of the Earth* (1695), was, unsurprisingly, focused on those fossils that were most clearly of organic origin. He interpreted them all as having lived in an antediluvial world. But he claimed that the Flood, which he imagined as an extremely violent event, had destroyed that world entirely. Universal gravitation—Newton's still-novel idea—had been temporarily suspended; and the materials of the Earth, other than the organic, had been churned up into a kind of soup or thick suspension. Then, with the resumption of gravity, these materials had settled out to form all the successive rock strata that are widely visible in cliffs and quarries. Fossils alone survived to give evidence of the world before this catastrophe, but they could do so only indirectly. For Woodward argued that they were not preserved in the places where the creatures had lived, nor even with any trace of those original habitats, but only where they had ended up after settling out of the diluvial soup (in modern terms, they were

FIG. 2.8 Scheuchzer's engraving of "A man, a witness of the Flood and a divine messenger" (*Homo diluvii testis et theoskopos*, 1725). As a trained physician he should surely have recognized that this fossil, whatever it was, was certainly not human; his scientific judgment was perhaps led astray by his uncritical adoption of Woodward's interpretation of *all* fossils as relics of the biblical Flood. (A century later this fossil was identified, by the then leading comparative anatomist Georges Cuvier, as a giant salamander, an extinct amphibian).

all "derived" or "reworked" fossils). Like his predecessors, Woodward took it for granted that all this had taken place within the chronologists' short timescale. No greater span of time was needed.

In the early 18th century many other naturalists, whether or not they adopted Woodward's speculative ideas about the character and cause of the Flood, followed his lead in claiming that fossils now provided an overwhelming case for its historical reality. Among them was the Swiss physician Johann Scheuchzer, who published a Latin translation of Woodward's book, making it accessible internationally. Like Woodward, Scheuchzer in his own prolific publications attributed all fossils of organic origin to the biblical mega-event. Half a century earlier, Kircher had illustrated his textual commentary on the Flood story with a pictorial inventory of all the living animals that must have crowded into Noah's Ark. Scheuchzer, at the equivalent point in a rather similar compilation, inserted a pictorial account not of living animals but of his own fine fossil collection. It was a significant change. Fossils, as natural antiquities that were relics of the Flood, had now become central to the ongoing debate about the Earth's own history. Scheuchzer even claimed that one of his specimens was the skeleton of "a man a witness of the Flood and a divine messenger" serving to warn his own generation of the historical reality of that dire event.

He convinced himself that this unique fossil supplied the decisive evidence that had previously been missing: evidence that the event that had entombed the remains of so many plants and animals was indeed the biblical Flood in which Noah's contemporaries had also perished.

This diluvial explanation of fossils did not go unchallenged. It certainly entailed a far from literal interpretation of the narrative in Genesis. Had a Flood that lasted a mere biblical "forty days" been long enough to deposit all the thick layers of sediment (later consolidated into rock strata) and to embed in them all these shellfish and other fossils? Or, if they had been swept suddenly and violently up on to the land by some kind of brief mega-tsunami, could that be reconciled with the biblical account of a rise and fall in sea level that had been equable enough for Noah's Ark to ride out its momentous voyage unscathed? Such questions pointed inescapably—not for the first time—to the necessity of critical biblical interpretation.

It would be easy in retrospect to dismiss Woodward's work, and that of Scheuchzer and others who were inspired by it, as worthless products of a foolish obsession with proving the historical reality of the biblical Flood. But their diluvial theory did help consolidate the still-novel idea that the Earth had had a true physical *history* that could be reconstructed from the material evidence of its natural antiquities. It also focused the attention of naturalists on fossils, which later opened up investigation of the Earth's history in an unexpectedly fruitful way.

## PLOTTING THE EARTH'S HISTORY

At the time, however, there was little about the natural world, apart from the unique mega-event of the Flood, to suggest that the Earth and its living beings had had any eventful *history* at all. Even in the early 18th century, any sense that there might have been a distinctive sequence of natural events or periods *before* the Flood continued to find more support and inspiration in the Creation narrative than in any fossil evidence. A telling example was Scheuchzer's "hexameral" sequence of pictorial scenes, depicting the six "days" of the first Creation story. It was published near the start of his huge illustrated commentary on biblical history (*Physica Sacra*, 1731–35; "physics" was very broad in meaning at this time), which made full use of his wide

TAB. XXII.

GENESIS Cap. I. v. 24. 25.
Opus sextæ Diei.

I. Buch Mosis Cap. I.v. 24. 25.
Sechstes Tagwerck.

**FIG. 2.9** The story of Creation visualized: one of Scheuchzer's imagined historical scenes, from the sequence published near the start of *Physica Sacra* (1731–35), his multi-volume illustrated commentary on the Bible. This engraving is of "the work of the sixth day" and shows the world just before the creation of Adam; the caption is in Latin and German, to cater respectively to an international and a more local German-speaking readership. The scene is shown like a painting, within an ornate Baroque frame; it is, as it were, a picture of the past as it might have been depicted by a time-traveling naturalist. All Scheuchzer's pictures adopted established artistic conventions for scenes from sacred or secular history, in this case scenes of the Garden of Eden. The animals and plants are of *living* species, but this kind of picture later provided a model for imagined scenes of the Earth's much deeper history, the inhabitants of which might have been quite different from any known alive.

scientific knowledge. These pictures offered imagined views of the world at many successive moments in the "week" of primal Creation (the bulk of the work depicted much later events, from the Flood onwards). In the course of the "third day," for example, a lifeless landscape was followed by one with mature trees and other plants of familiar living kinds. Later, in the course of the "sixth day," a scene of the Garden of Eden newly stocked with assorted animals, again of living kinds, was followed by one in which Adam had arrived to take charge. Did Scheuchzer really think that these had been literally days of twenty-four hours apiece? He may well have followed other commentators in inferring—as the well-established principles of biblical interpretation allowed—that the "days" could well denote periods of greater and perhaps indefinite length. Yet Scheuchzer's magnificent scenes incorporated no visual allusion whatever to his own great collection of fossils, as possible evidence for these remote periods: as already mentioned, all his fossils were deployed instead in his account of the Flood, later in the work.

Irrespective of the timescale, what is most significant about Scheuchzer's scenes, like the Genesis narrative that inspired them, is that they formed a coherent *sequence*, starting with a lifeless world and displaying the successive addition of plant life, marine life, the higher forms of terrestrial life, and finally human life. These scenes reinforced visually the sense that the natural world must have had an intelligible *history* of its own, even if it was imagined as no more than a brief prelude to a lengthy human history. At its start this history had been pre-human, and its very earliest phase had even been pre-life. But the idea of such a sequence, and the evidence for it, were drawn as yet almost exclusively from the narrative in Genesis, not from the natural world itself.

However, this was an outline of the Earth's total history that could readily be expanded to fill a much longer timescale, if and when the natural evidence was found to require it. But in the late 17th and early 18th centuries, on which this chapter has been focused, there was little that seemed to demand any such lengthening of the traditional short timescale that chronologists such as Ussher had so carefully quantified. Only on the margins of these debates, and only occasionally, did a few savants express some doubt about whether a few millennia were sufficient time to account for everything that was now becoming known. For example, the great English naturalist John Ray, convinced that many fossils were indeed truly organic but dissatisfied by Woodward's diluvial explanation of them, commented in a letter to another savant that from this there might follow "such a train of consequences, as seem to shock the Scripture-History of the novity [newness] of the World." Yet if he was hesitant to go down that path, it was precisely because it seemed to entail questioning the reliability of scripture *as history*, which he and most of his contemporaries took for granted as common sense. Hooke, less troubled by such doubts, suggested that natural antiquities such as fossil shells might turn out to be even older than the most ancient of human antiquities. Yet it is not clear that even he thought they would extend the reach of history beyond the earliest or "mythical" age of humankind.

One of the few examples of speculation about a possibly far longer timescale was that of the English savant Edmund Halley (now most famous for calculating the orbit, and accurately predicting the return, of the comet that bears his name). He tried to work out a method to calculate the total age of the Earth, by estimating the rate at which the world's rivers are currently adding to the salt content of the oceans; he concluded that "perhaps by it the World may be found much older than many have hitherto imagined." Yet his explicit goal in his paper on this point, read at the Royal Society in London, was to repudiate any claim that the Earth was eternal, by proving that its history must have had a beginning at some point in time, however remote (like many others he suggested that the "days" of Creation could have been lengthy periods of time). For most savants around the end of the 17th century and the start of the 18th, the real threat was an eternal and uncreated world, not a long timescale.

The debates outlined briefly in this chapter, although focused particularly on fossils, had a bearing on much broader issues. These were incorporated into a much more ambitious kind of theorizing, which must be traced (in the next chapter) from the 17th through to the end of the 18th century, before this narrative can return to its main theme of the detailed reconstruction of the Earth's own history.

# 3

## Sketching Big Pictures

If fossils—using the word from now on in its modern sense—
were truly nature's own antiquities, they could supplement human documents and other antiquities and shed further light
on the earliest phases of human history and its physical environment. This conclusion, as argued by Steno, Hooke, and
many other savants in the late 17th and early 18th centuries, was
matched at the same time by a quite different way of studying the
Earth. Rather than piecing together the sequence of events that
marked the Earth's *history*, they could try to work out the underlying *causes* of these events. Rather than borrowing concepts
and methods from chronologists, antiquaries, and other historians, they could draw on the work of natural philosophers (in
this context roughly equivalent, in modern terms, to physicists),
and try to apply the basic "laws of nature" to the Earth's physical
features. In principle these two projects were complementary.
To claim, for example, that the Flood was a real historical event,
recorded not only in a scriptural text but also by nature's own
antiquities, was quite compatible with trying to work out how
such a dramatic physical event might have been caused. Steno
and Hooke were tackling both when they suggested physical

causes for the occurrence of fossil sea shells high and dry on land. But in fact the attempt to understand the Earth in terms of natural causes developed into a kind of theorizing that was distinct from the attempt to reconstruct its history.

Those who proposed causal theories often aspired to construct a Big Picture, as it were, that would explain the Earth as a whole; and not just the past that had led to its present state, but also the future that necessarily lay ahead, given the continuing operation of nature's unchanging laws. A model for such scenarios of both past and future came from a well-known work published earlier in the 17th century. René Descartes' *Principles of [Natural] Philosophy* (*Principia Philosophiae*, 1644) had outlined a conjectural picture of the whole cosmos, governed throughout by what the French philosopher claimed were the fundamental laws of nature. Within this grand vision he had analyzed the Earth: no longer as a unique body at the very center of the universe, as it had been traditionally, but as one of any number of similar planets that were likely to be orbiting stars widely scattered throughout space (the possibility of such a *"plurality of worlds"*— perhaps inhabited like ours—was hotly debated long before the modern era of space exploration and SETI). In the perspective of the new astronomy, the Earth was unique only in being the most accessible body of its kind. Descartes had argued that *any* Earth-like body, anywhere in the universe, had already undergone, or would in the future undergo, a similar sequence of changes that were built in from the start.

This sequence was determined or programmed, as it were, by the body's initial state (supposedly as a former star) and then by the laws of nature acting upon it. In Descartes' view its structure would therefore necessarily change over time in a predictable way. From its initial state as a ball of incandescent matter, it would gradually separate out into various concentric layers of differing composition. Among these would be a crust or outermost solid layer. At some point in time, he argued, this crust would crack and break up. Some parts would then collapse into an underlying liquid layer, while others would buckle upwards into an overlying gaseous layer. In the specific case of our own Earth, this would produce what could be recognized as a varied topography of mountains, continents, and oceans, enveloped by an atmosphere above and with an unseen core (and a hypothetical liquid layer) beneath.

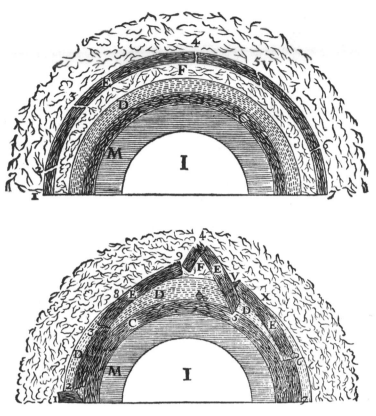

FIG. 3.1 Descartes' diagrammatic slices or sections through any Earth-like body at two successive phases in its predictable sequence of changes, before and after a solid crust (E, the outer dark layer) breaks up. Parts of the crust buckle upwards into an outermost gaseous envelope while other parts collapse into an underlying liquid layer (D). This produces a body with an irregular surface topography, and with an atmosphere above and an unseen core beneath. (Only half the sphere is shown at each phase, to save engraving costs and space on the page of his *Principia Philosophiae*, 1644.)

Descartes did not specify the kind of timescale on which he thought Earth-like bodies would change in this way. He did not need to. All that mattered was that some such sequence was necessitated by the laws of nature, whatever the rates at which the natural processes were operating. But in the tense political world of the Counter-Reformation his vagueness about the timescale was also prudent. Galileo had notoriously tangled with the Catholic authorities in Rome over the wider implications of his cosmology. Descartes had no wish to court a similar fate. But in fact his theory, when applied to our own planet, could easily fit into the few millennia computed by the chronologists; and he himself, like Steno, Hooke, and most other savants, may have seen no compelling reason to question this (the timescale for the universe as a whole might be quite another matter).

Anyway, during the rest of the 17th century and through much of the 18th, Descartes' well-known theory was used as a model for many others with the same focus on the laws of nature that might have governed the physical development of the Earth. At least in principle, any such theory would try to account not just for specific phenomena (say, earthquakes or volcanoes) but for *all* the major physical features and natural processes found on the Earth. It would give a causal explanation of the whole physical "*system*" on which the Earth works, in past, present, and future (it is no accident that its nearest modern equivalent, "Earth-systems science," uses the same key word). "*Theory of the Earth*," as it was called, became a distinctive scientific genre, in the same sense that novels, sonnets, landscapes, and symphonies are literary and artistic genres.

## A "SACRED" THEORY?

The first major work to use this title—with one significant additional word—was the *Sacred Theory of the Earth* (*Telluris Theoria Sacra*, 1680–89) published by the English savant Thomas Burnet, who was as much at the center of the intellectual life of his time as Ussher had been half a century earlier. Burnet was not a fundamentalist in the modern sense, but he certainly was concerned to integrate what had long been regarded as the two complementary sources of trustworthy human knowledge, nature and scripture, God's "Works" and God's "Word." He therefore referred both to the unfolding of physical events under the unchanging laws of nature, and to the testimony of scripture about the recorded past and the prophesied future. His theory was summarized revealingly in the frontispiece of his volumes. It depicted the Earth with a finite past and future, with the present at mid-point, and with the cosmic Christ presiding symbolically over the whole drama from beginning to end, Alpha to Omega.

Descartes' initially unbroken crustal layer was identified as the smooth perfection of the primal human world of the Garden of Eden. The breaking up of the crust later produced a global Flood, identified of

**FIG. 3.2** Burnet's visual summary of his *Sacred Theory of the Earth*: the frontispiece of its English translation (1684). The chronologists' finite linear history is here curled into a circle to symbolize its future completion: Christ, with the self-description attributed to him, "I am the Alpha and the Omega" in Greek, stands astride the first and last of seven phases. In clockwise order, an initial Chaos is followed by the smooth perfection of a paradisal antediluvial world, and then by the global Flood (showing a tiny Noah's Ark afloat). The present world, with familiar continents and oceans, is at mid-point. Still in the future—but predictable, given the further operation of the unchanging laws of nature—are the global volcanic "Conflagration" that will destroy the present world, the subsequent perfection of the Earth during Christ's millennial reign, and the final End that will turn the Earth into a star. Past and future are strikingly symmetrical, either side of the present. Observing the sequence from outside its space-time framework are angels, representing the eternity of the divine realm. But the Earth itself is not eternal: the circular sequence has a clear beginning and end.

course as Noah's. When the waters subsided, they revealed the broken and imperfect world of the present, with its irregular geography of continents and oceans. Moving into the future, the further working of the natural laws that produced the Flood would in due course produce worldwide volcanic mega-eruptions, identified as the fiery apocalypse supposedly predicted in the Bible. This would purge the world and make it smooth and perfect for a second time, fit for Christ's future millennial earthly reign. Finally, and again by the continuing operation of the laws of nature, the Earth would be transformed into a star. This entire sequence lay tacitly within the chronologists' kind of timescale (the penultimate period, the "millennium" itself, was expected to last literally 1000 years). Burnet explicitly rejected any suggestion that the world might be eternal or have a cyclic kind of history. The circularity of his sequence expressed its completeness in the "Alpha and Omega" of Christ: it was not just one cycle in an infinite succession of similar cycles, as eternalists would have portrayed it.

Burnet's theory was immensely influential, and not only among savants. Ironically, he found himself accused of atheism, despite the "Sacred" reference in his title and his scholarly treatment of the scriptural evidence, and despite his explicit rejection of eternalism. He was also criticized for ignoring important scientific evidence. For example, in the biblical narrative the primal humans were ejected from the Garden of Eden into a world so "fallen" that their descendants (other than Noah and his family) had later deserved to perish in the Flood. In contrast, Burnet ignored the Fall and depicted the whole antediluvial period as paradisal and perfect. This original perfection entailed the total absence of any seas (which in the Bible often symbolized nature as chaotic). So on his theory there was no obvious way that marine fossils could have been embedded in rock strata, and in fact he ignored the lively debate about rocks and fossils that was engaging his contemporaries at the Royal Society and elsewhere. Much more seriously, Burnet attributed both Flood and Conflagration solely to the operation of natural laws, which seemed to make these mega-events determined, or programmed in advance, and therefore in principle predictable. This was difficult to reconcile with their traditional interpretation as marks of divine judgment on the unpredictable moral behavior of fallen human beings.

However, Burnet's theory was taken very seriously, not only by its critics but also by those who found it persuasive. Isaac Newton, for example, suggested to Burnet how it could be improved: with an initial difference in the Earth's speed of rotation, the "days" of Creation might in fact have been years (though this hardly extended the timescale!). Newton's admirer William Whiston, later his successor at Cambridge, published *A New Theory of the Earth* (1696) in which he replaced Descartes' laws of nature with Newton's. He claimed that this improved on Burnet's theory and brought it up to date. Specifically, he suggested that *comets* were the likely physical cause of both the Flood in the past and the Conflagration in the future (comets were better understood in the light of Newton's work, but were thought to be massive bodies that might well be capable of such large effects). But Whiston's main objective was the same as Burnet's: to show, as he put it in his subtitle, that the scriptural history was "perfectly agreeable to Reason and Philosophy." Religion and natural science were claimed as compatible, and certainly not in intrinsic conflict, though the relation between them remained a matter of vigorous debate.

"Theory of the Earth" was now well launched as a scientific genre, although a highly controversial one. In fact, Burnet's work prompted such a profusion of books and pamphlets—many of them claiming to put forward the one and only *true* Theory—that one critic referred scornfully to the whole project as absurdly speculative "worldmaking." Nevertheless, it continued to be popular among savants throughout the 18th century. But it was modified in two important ways. First, it absorbed almost effortlessly the vastly enlarged sense of the Earth's likely timescale that emerged during that century (since the sources for this were quite distinct from the genre itself, it will be described in the next chapter). Second, in the cultural climate of the intellectual movement of the Enlightenment, "Theory of the Earth" was generally narrowed down into a project concerned exclusively with the laws of nature in the physical world. The attempt to integrate the evidence of nature and scripture, as seen in Burnet's theory and many others, was generally abandoned or at least marginalized. The divine dimension was rarely repudiated in explicit atheism. But the theology of "*deism*," which was widely adopted by Enlightenment savants, relegated it to the margins. In contrast to traditional Christian (and Jewish) "*theism*," with its dynamic understanding of God as utterly transcendent yet interacting with the world in the course of human history, deists proposed that a "Supreme Being"—treated in practice as almost impersonal—had designed and created the universe in the first place but thereafter left it to run on its own. In the perspective of deism, the alleged physical effects of the biblical Flood were generally played down or denied altogether, while the Creation story was usually dismissed as scientifically worthless. This had a profound effect on the study of the Earth. It usefully focused attention on the timeless laws of nature that govern the Earth's *causal* processes. Conversely, however, by rejecting any evidence from scripture it also diverted attention away from the Earth's possibly unrepeated and contingent *history*. This will be illustrated here by the examples of two major Enlightenment savants, and by a third whose work in contrast tried to reinstate a role for scripture and therefore also for a truly historical study of the Earth. Each of these Big Pictures left an influential legacy to the 19th century and even beyond it.

## A SLOWLY COOLING EARTH?

One of these grand theories was due to Georges Leclerc, count Buffon, who for several decades was the director of the royal natural history museum and botanic garden in Paris (now the Muséum National d'Histoire Naturelle, in the Jardin des Plantes). He was a powerful figure at the heart of French cultural and political life. His huge multivolume work on *Natural History* (*Histoire Naturelle*, 1749–89) was designed as a comprehensive survey of all three "kingdoms" of the natural world, "animal, vegetable, and mineral," though as it turned out it dealt mostly with animals. This was "natural history" in its traditional sense of a static description of nature, not a "history" of nature in the modern sense of a narrative of changes extending through time.

One of Buffon's introductory essays sketched a theory of the Earth, regarded as the environment of the organisms he was planning to describe. He portrayed the Earth as a scene of unceasing gradual change, but change without any overall direction. The physical causes of change in the past were, he argued, simply those—such as erosion and deposition—that could be observed in the present and could be expected to continue in the future. In some places the sea was seen to be eroding the land; in others, new land was being formed and the sea was being displaced. At different times every part of the globe would have been, or would be in the future, both land and sea. Buffon's theory portrayed an Earth in a *"steady state"* of dynamic equilibrium. It was therefore an Earth without any real history. Significantly, his theory lacked any reference either to a primal Creation or to a subsequent Flood (elsewhere he claimed, disingenuously, that the Flood would have left no physical traces because it was a miracle). If the Earth was a scene of constant but non-directional change, its total timescale was irrelevant and could be left undefined. But then the Earth might be thought to have no beginning or end, existing uncreated from and to eternity. This was one of the points—others did not touch on Buffon's ideas about the Earth—on which his work was criticized by some of the Paris theologians (the Jansenists), although others (the Jesuits) were more positive about it. As with Galileo in Rome a century earlier, "The Church," even its Catholic branch, did not speak with one voice. But Buffon was much too powerful in royal circles to be formally censured, though he did issue an anodyne statement of religious orthodoxy, conceding that his scientific ideas were merely hypothetical (as indeed they were).

Anyway Buffon soon published another essay about the origin of the Earth, which undercut any suspicion that he was a covert eternalist: if the Earth had had an origin it could not be eternal (the universe, as usual, might be quite another matter). He suggested that at some point in the past a massive comet, in a close encounter with the Sun, had torn off a plume of incandescent material that had then condensed into a string of planets, of which the Earth was one. Buffon's theory, like Whiston's, was inspired by Newton's much admired natural philosophy, which helped to make it scientifically respectable (Buffon translated some of Newton's work into French, which was gradually replacing Latin as the main international language of all the sciences).

Buffon had now outlined two contrasting theories about the Earth: a steady-state theory based on its present processes, and a theory about its sudden origin at some distant point in past time. Many years later he combined the two in a book-length essay *On Nature's Epochs* (*Des Époques de la Nature*, 1778), published in one of the last volumes of his huge work. Much had happened in the meantime to make this new theory of the Earth seem quite plausible. Measurements of the rising temperature in deep mines (in modern terms the "geothermal gradient") had confirmed the reality of the Earth's internal heat, which Buffon thought could best be explained as heat left over from the incandescent origin that he and others had already suggested. Accurate measurements made on scientific expeditions to Lapland and Peru had proved that the Earth's overall shape is an oblate spheroid, as Newton's laws predicted if it had once been a rotating fluid body. Fieldwork around Europe had confirmed Steno's inference that the rocks lowest in position contained no fossils at all, so they might date from before any life existed. Rocks overlying them and obviously younger contained many strange fossils; some, such as giant ammonites, looked distinctly tropical. In loose deposits, overlying all the solid rocks, were the bones of elephants and rhinoceros, yet they were found even in the north of Siberia. And there were no traces of human fossils anywhere (discounting Scheuchzer's dubious "man a witness of the Flood").

Although Buffon himself had not contributed directly to any of this research, such conclusions were well known to him from what was reported and discussed at the Académie des Sciences in Paris and from the fossils acquired for his museum. They suggested to him that the Earth might have cooled progressively from the extremely hot origin he had already proposed, and that the successive phases of this grad-

As in civil history title deeds are consulted, coins are studied, and ancient inscriptions are deciphered in order to determine the epochs of human revolutions and to fix the dates of human events; so also in natural history it is necessary to excavate the world's archives, to extract ancient monuments from the Earth's entrails, to collect their remains, and to assemble in a body of evidence all the marks of physical changes that are able to take us back to the different ages of nature. This is the only way to fix some points in the immensity of space, and to place a certain number of milestones on the eternal road of time.

FIG. 3.3  The opening words of Buffon's *Nature's Epochs* (*Époques de la Nature*, 1778), in which he staked his claim to be reconstructing the *history* of the Earth by following the same methods as historians of the *human* world. Traditional descriptive "natural history" was to be turned into nature's own dynamic "history" in the modern sense. His reference to "milestones on the eternal road of time" does not imply that he thought them equally spaced, and what was eternal was just the abstract dimension of time, not the chronology of real events in history.

ual cooling could be reconstructed. Significantly, his title borrowed a key word from the chronologists. They had defined the major turning-points in human history as its "*epochs.*" Buffon set out to reconstruct the sequence of *nature's* epochs. Like Hooke a century earlier, he also borrowed other key words from the antiquaries. Finds such as fossils were nature's "monuments," relics surviving from the past. Analogies with coins and inscriptions, documents and archives, made it even clearer that he was claiming to reconstruct the Earth's own *history*.

Buffon outlined his conjectural history of the Earth by defining six epochs, ranging from the Earth's origin as a rotating hot fluid body—a Fireball Earth, as it were—to the appearance of large tropical land animals even at high latitudes. (The origin of new forms of life was no great puzzle for Buffon and his contemporaries; they attributed it to some kind of natural process of "spontaneous generation," not to evolution from other living forms nor to direct divine action.) The parallel with the six "days" of Creation could not be missed, though Buffon may have intended it as a sly parody rather than a respectful reinterpretation of Genesis. But anyway it underlined the directional character of his narrative: this was a major change from his earlier steady-state theory. He had planned to put the first appearance of humans at the sixth epoch (as in the sixth "day" in Genesis). But this would have made them contemporary with the large fossil mammals, of which he suspected that at least one species was extinct. So just as his work was about to be published he added a seventh epoch to mark the first appearance of humans: this kept it safely in its traditional place as the final major event in the story (though it also put humans into the supreme position occupied in Genesis by the divine Sabbath rest!). More importantly, however, Buffon's last-minute change made explicit what other savants were already taking for granted: that almost all of this history of the Earth (and of living things) was pre-human. Humans might still be the culmination of the story, but its non-human prelude was now hugely expanded. To put it another way, human history was reduced to the final scene in a much longer drama.

Just how much longer was another issue on which Buffon made explicit what other savants were already taking tacitly for granted, namely a total timescale far beyond the traditional few millennia. Unlike others, however, Buffon tried to put accurate quantitative figures to it. Using the forge on his rural estate, he timed the rate of cooling of small balls of different sizes and various materials, from white heat to room temperature, and then scaled the results up to the size of the Earth. This gave him a total age of the Earth of about 75,000 years, though he suspected this was a gross underestimate and he speculated privately with figures up to ten million. Even his lower figure, which he published, knocked the bottom out of the chronologists' usual calculations, so he certainly did not keep his higher estimates to himself for fear of criticism by church authorities. Rather, he thought he had experimental evidence for the lower figure, but only a mere hunch for any higher one. What he did fear—correctly—was criticism from other savants for being too speculative: all his figures depended on the way he scaled up from tiny models to the real Earth, and of course on the validity of his cooling theory itself.

Buffon embedded his figure for the age of the Earth in a context that reveals something far more important about his theory. His whole sequence was based on the rate of cooling of an initially incandescent body, so it was equally applicable to all the bodies in the Solar System, both planets and satellites. Their rates of cooling would depend primarily on the size of each body, though also on its proximity to the heat of the Sun. In effect, Buffon's theory was modeled on Descartes' much earlier one: the sequence of events was strictly determined throughout—or, as it were, programmed in advance—depending only on the initial conditions of an incandescent state combined with the physical laws of cooling bodies. And this would apply not only to the past of each body, but equally to its future. In the case of the Earth, Buffon predicted and even calculated a date for the final annihilation of all its life, as it cooled further and the frozen wastes of the Arctic encroached on the rest of the globe, turning it into a Snowball Earth (to borrow a much later term). This shows how, for all his rhetorical use of metaphors such as nature's coins and inscriptions, epochs and monuments, Buffon's theory of the Earth was historical only in a very limited sense. It reconstructed the predictable course of the Earth's physical development through time, as a scenario of past and future, from Fireball to Snowball, under the constant laws of nature. But it lacked the messy unpredictable contingency of human history. It was in effect a secularized version of the Creation narrative in Genesis, adopting its emphasis on directional change but discarding the deep contingency derived from its foundation in divine initiative.

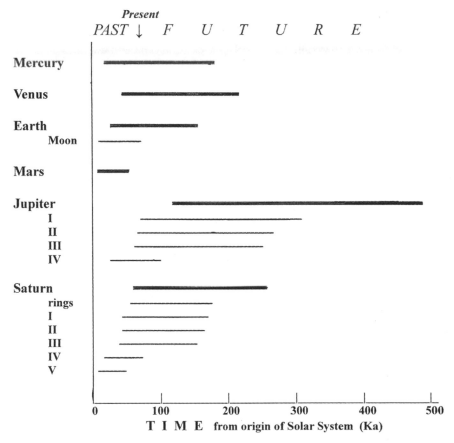

**FIG. 3.4** Buffon's calculations for the duration of life on each of the planets and their satellites, in thousands of years (Ka) from the initial formation of the Solar System. This diagram, drawn in modern style with time flowing from left to right, is based on Buffon's figures (1775), which were derived from his experiments with the rate of cooling of small model balls from white heat, scaled up to the real size of each body. He thought that life would be "spontaneously" generated on each, as soon as its surface was cool enough "to be touched," and would cease when the surface temperature reached the freezing point of water (he also assumed that all these bodies were solid and physically Earth-like). No distinction was made between past and future, and the position of the present could only be inferred from one figure for the Earth—at 74,832 years—in the middle of a dense matrix of other figures. It is striking that the anticipated future is far longer than the past. Buffon conceded that all his calculations were "hypothetical," but they did under-line his view that the course of development of all these bodies, and in particular the Earth, was determined and predictable, given the universal physical laws of cooling bodies.

Buffon's new theory of the Earth had a mixed reception. Again there were little local difficulties with the theologians in Paris, but they were even more muted than before. In the cultural capital of the Enlightenment such criticisms were in effect brushed aside as an irrelevance. The eloquent literary style of *Nature's Epochs* was much admired by its general readers, but savants tended to dismiss Buffon's essay as a mere "novel" because it was so speculative. Where it was based on concrete observations, they were largely those made by others, not by Buffon himself (his cooling experiments were an exception). He had done little or no fieldwork, which was now regarded as an essential qualification for any serious work of this kind. His all-explanatory system did help his readers to imagine the Earth as having possibly had a vast and varied past, stretching back far beyond even the earliest human history. But it remained for others to show that this immense and impressive panorama was anything more than an imaginative piece of science fiction.

## A CYCLIC WORLD-MACHINE?

Only a few years later, another quite different theory of the Earth was added to the crowd of such conjectural "systems." It was due to the Scottish savant James Hutton, who belonged to the intellectual circle in Edinburgh that included famous Enlightenment figures such as David Hume and Adam Smith. Like them, Hutton regarded himself primarily as a philosopher. His most massive work was on epistemology: *An Investigation of the Principles of Knowledge* (1794) was as wide in scope as its title suggests. His *Theory of the Earth* (published in outline in 1788 and more fully in 1795) was just one component in a much more ambitious intellectual project. In thinking about the Earth, Hutton, like Buffon, took it for granted that time was available to nature, as it were, in as large amounts as were necessary to achieve the required effects; to put it another way, a philosopher was entitled to invoke as much time as was necessary to explain what could be observed. Again like Buffon, Hutton took it for granted that such explanations should be in terms of the generally slow natural processes—such as erosion and deposition—that could be observed in action in the world around him. Both these principles were already widely adopted among savants in the later 18th century, and Hutton was not the first to put them to creative use (a mistaken belief that he was, spiced with Anglophone or even Scottish chauvinism, has given him an undeserved modern reputation as geology's uniquely important founding "Father"). Hutton did not mention Buffon's work, though he must have known of it: Buffon was a towering figure on the international scientific stage, and Hutton—like any educated Briton at this time—could read French with ease.

Hutton had earned a medical degree at the great Dutch university of Leiden by writing a dissertation (in Latin) on the circulation of the blood in the human body. Later, back in Scotland, he contributed to discussion of what is now called the hydrological cycle: water falls as rain and flows in rivers to the sea, where evaporation forms clouds that drop rain over the land and so complete the circulation. Such cyclic or steady-state systems were much in vogue among Enlightenment savants, as ways of understanding things in both the natural and the human worlds. So it is not surprising that Hutton, like the early Buffon, formulated yet another steady-state cyclic system for the Earth itself.

Hutton reasoned that human life depends on the life of animals and plants, which in turn depend on the soil (he owned farms near Edinburgh and had thought long and hard about agriculture). Soil is formed by the breakdown of the underlying bedrock, but it is continually being washed away into rivers and then into the sea. Hutton argued that the land must therefore disappear in the long run, by being eroded down to sea level, making it no longer available to support human life. But there might be some other process that creates new land areas to replace those lost. The material worn down from the land and swept into the sea must end up being deposited on the sea floor. If it was consolidated there, forming new rocks, and the Earth's crust in that part of the globe was then slowly heaved up above sea level, new land would be formed and the cycle could be repeated. Hutton claimed that this essential "renovating" process must be powered by the huge expansive force of heat in the Earth's deep interior; it was this that made the Earth's surface a scene of perpetual change.

The ultimate purpose of this dynamic but steady-state "system of the habitable Earth"—as Hutton revealingly called it, when in 1785 his preliminary paper was read at Edinburgh's new Royal Society—was to ensure that the Earth remains permanently habitable by humans, from and to eternity. His theory was based on his natural theology, and specifically on his deistic belief that the Earth had been constructed to support rational beings who can appreciate its intelligent and beneficent design (the modern creationist proponents of Intelligent Design are merely rewarming an ancient concept). It was, as Hutton put it, "a system in which wisdom and benevolence conduct the endless order of a changing world"; "what a comfort for man," he added, "for whom that system was contrived, as the only living being on this earth who can perceive it." This kind of language was not just a politically prudent cover for atheism (analogous to the Marxist language that sometimes disfigured scientific work published under the Soviet regime). It pervaded Hutton's writing and made no sense except within the framework of his deistic beliefs, of which it was a sincere expression.

Only after presenting his steady-state theory of the Earth in public—or at least, to other Edinburgh savants—did Hutton do extensive fieldwork around Scotland, looking for what he expected to find if his theory was valid (he was using the scientific method now called "hypothetico-deductive"). He duly found evidence that granite, the distinctive rock usually lowest in position, could not really be the oldest of all, because it seemed to have been squirted as an intensely hot liquid into cracks in overlying rocks, before cooling into a crystalline solid. He took this to be evidence that beneath the Earth's crust was an extremely hot fluid interior, which could power the upheaval of the crust to form new land areas. He claimed that this kind of upheaval had happened again and again. He called the Earth a "machine," in allusion to the steam-engines that were such a spectacular feature of the early Industrial Revolution in Britain. Steam-engines demonstrated the huge expansive power of heat, as one phase in a repeated cycle that could continue to operate indefinitely. This cyclic character, Hutton argued, was just what made the Earth a natural machine analogous to a steam-engine.

Hutton also looked for, and found, further evidence for the cyclic operation of this natural machine, in places where rocks that had been formed in its successive cycles were in contact. If one set of strata, deposited as horizontal layers on some sea floor long ago, was subsequently heaved up to form dry land, then eroded down to sea level by rain and rivers, then covered by a second set of strata deposited on a

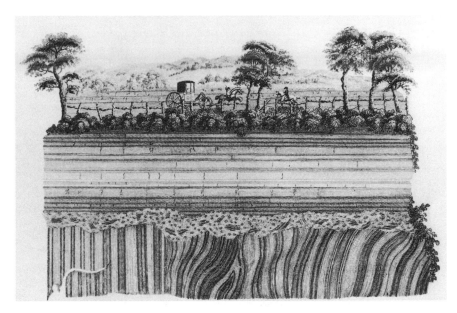

FIG. 3.5 Hutton's published engraving (1795) of the angular "junction" (in modern terms, a major *unconformity*) between two sets of strata, which he found in 1787 in a river gorge at Jedburgh in southern Scotland. The lower and older strata, originally deposited as horizontal layers, have been heaved up into a vertical position; then eroded and the strata truncated (with some fragments preserved above); then covered by an upper and younger set of strata, which has since been elevated to form the present dry land, with plants, animals, and human life. In Hutton's view these rocks represented two successive cycles of deposition in an ocean, followed by upheaval to form new land, followed by its erosion: they were relics of two successive habitable "worlds."

later sea floor, and again heaved up into dry land, this could be taken as evidence for at least two successive "habitable worlds." Hutton saw no reason to doubt that there had been others before them, and that there would be others to come in the future. To use his own favorite analogy, the Earth's "system" was just as repetitive as the orbiting of the planets in the Solar System; successive "worlds" were no more distinctive than successive orbits. He did not think that fossils suggested otherwise. On this point all that mattered to him was that plant and animal fossils proved the existence of both land and sea in former "worlds." Hutton conceded that there was no fossil evidence of *human* life before the times of recorded human history, but he treated plant and animal fossils as surrogates or proxies for that missing evidence. In his deistic system of intelligent design, any "world" with lots of

> WE have now got to the end of our reasoning; we have
> no data further to conclude immediately from that which ac-
> tually is: But we have got enough; we have the satisfaction
> to find, that in nature there is wisdom, system, and consistency.
> For having, in the natural history of this earth, seen a succession
> of worlds, we may from this conclude that there is a system in
> nature; in like manner as, from seeing revolutions of the pla-
> nets, it is concluded, that there is a system by which they are
> intended to continue those revolutions. But if the succession
> of worlds is established in the system of nature, it is in vain to
> look for any thing higher in the origin of the earth. The result,
> therefore, of our present enquiry is, that we find no vestige of a
> beginning,—no prospect of an end.

FIG. 3.6 The last paragraph of Hutton's *Theory of the Earth* (1795), with its famous final sentence claiming that the Earth's "system" shows no sign of either beginning or end. It summarizes Hutton's steady-state eternalistic theory, in which there was a "succession of worlds"—here explicitly analogous to the successive orbits of the planets around the Sun—without origin and extending indefinitely from past into future. The language of "wisdom" and "intention," and indeed "system" itself, expresses Hutton's deistic theology: the Intelligent Design of the Earth-machine enables it to provide human life with habitable land from and to eternity. (The *f*-like form of the letter *s* was common in printed material at this time.)

non-human life would have been pointless, unless in reality humans were also present to fulfil its ultimate purpose.

Hutton therefore saw no reason to think the Earth ever had been, or ever would be, significantly different from what it is like at present. Despite the continual erosion of the land and its disappearance beneath the sea, other dry land was always emerging elsewhere to re-place what was lost. So there had always been, and always would be, dry land available for human habitation. Hutton's steady-state Earth, as a "system" wisely designed to support human life from and to eternity, was therefore even less historical than Buffon's developmen-tal one. His successive "worlds" formed an endless sequence in time, but not a real *history* of the Earth, any more than the repeated orbits of the planets constituted a real history of the Solar System.

Hutton's theory was widely noticed by savants in his own country and across the rest of Europe. That it portrayed the Earth as eternal was crystal clear to his contemporaries, both supporters and crit-ics. For example, Erasmus Darwin (Charles's grandfather) noted ap-provingly that according to Hutton "the terraqueous globe has been, and will be, eternal." Another writer cited it in his own support in a work entitled unambiguously *The Eternity of the Universe*. On the other side, one reviewer noted derisively that Hutton claimed there had been "a regular succession of earths from all eternity! and that the succession will be repeated for ever!!" And a mineral surveyor—who knew a thing or two about rocks—complained that Hutton "warps and strains everything to support an unaccountable system, viz. the eternity of the world." This kind of criticism was directed in part at the scientific features of his theory, for example his claim that all soft sediments must be turned into hard rocks by being heated intensely or even melted on the ocean floor.

Hutton's system was not ignored or neglected; and living in one of the cultural centers of the Enlightenment it was of course unthinkable that he would have been persecuted for his opinions. But by the end of the 18th century the genre of "Theory of the Earth" was generally regarded by savants as having outlasted its usefulness. Hutton's example of it, like Buffon's, was widely considered too speculative to be taken seriously. Although some of his detailed observations were accepted as valuable, his theory might well have been forgotten, along with other 18th-century works in the same genre, had it not been repackaged, after Hutton's death, to suit the scientific tastes of a new century.

## WORLDS ANCIENT AND MODERN?

There was, however, one of Hutton's critics—and one of the most perceptive—whose work foreshadowed both the transformation of the genre and its demise. Jean-André Deluc (or de Luc) was a citizen of Geneva (a city-state not yet a part of Switzerland) who earned a fine reputation as a meteorologist and scientific instrument-maker. In his thirties he migrated to England, where he joined the Royal Society and was appointed intellectual mentor to Queen Charlotte, the German-born wife of King George III. During the rest of his long life he traveled widely in western Europe and published most of his work in his native French. Deluc regarded himself as an Enlightenment philosopher, no less than Buffon and Hutton, but unlike them he was not a deist, let alone an atheist. He called himself a "Christian philosopher" or theist. He was not a religious fundamentalist, but he did believe that scripture was a trustworthy guide to human life, and specifically that it embodied a reliable record of divine initiative: he took the Bible seriously as history. Like many before him, he was particularly concerned to demonstrate the reliability of the biblical narratives of Creation and the Flood, again as history (this has earned him a modern reputation as negative as Ussher's, but with even less justification).

Deluc's earlier work on these themes was published just after Buffon's *Epochs* and a few years before Hutton's *Theory*, but it offered an interpretation of the Earth radically different from either. His six volumes of *Letters on the History of the Earth and of Man* (*Lettres sur l'Histoire de la Terre et de l'Homme*, 1778–79) were addressed to his royal patron, who, being a highly intelligent woman, probably read them with close attention. At the start he suggested tentatively that this kind of theorizing about the Earth should be called "geology," by analogy with "cosmology" for the universe as a whole. The word, with some change of meaning, eventually stuck. In later years he elaborated his ideas in lengthy articles published in French, German, and English, in some of the scientific periodicals that were starting up all across Europe; his work was certainly well known to other savants. On the basis of his fieldwork around western Europe—far more extensive than Hutton's in Scotland, let alone Buffon's in France—Deluc described what he claimed was physical evidence for the historical reality of a major event in the Earth's recent history, which he identified with the biblical record of Noah's Flood.

Like Buffon and Hutton, Deluc based his argument on his study of proce sses observably active in the present, such as erosion and deposition; he called them *"present causes"* (*causes actuelles*, using "actual" in a sense still current in many European languages though now almost obsolete in English). To anticipate a later geological slogan, he believed that the present is the key to the past. Unlike Buffon, however, Deluc studied present causes at first hand, in the field; unlike Hutton, he claimed that they had not been active indefinitely in the places where they could now be observed. The field evidence, he claimed, showed that they had started acting on the present continents at a finite time in the relatively recent past. For example, major rivers such as the Rhine and the Rhône, laden with sediment eroded upstream, were forming deltas at their mouths, and these were growing at a rate that could be estimated from historical records. Deluc used the analogy of an hour-glass (a device more familiar in his day than in ours): at any moment the finite amount of sand that has trickled through shows that a finite amount of time has elapsed since the glass was last upended. Deltas were of finite size, so they likewise must have started being formed at a finite time in the past. Deluc later referred to such features as "*nature's chronometers*," alluding to John Harrison's extremely accurate marine clocks (the greatest technical achievement of the 18th century, which at last solved the longitude problem for navigation). Deluc's "chronometers" were far from accurate, but the analogy did enable him to argue that the start of what he called the "present world" could not be more than a few thousand years in the past (many of the features he analyzed would now be regarded as representing the few thousand years since the end of glacial or periglacial conditions in northern Europe at the end of the Ice Age).

Deluc claimed that this kind of approximate figure was quite adequate to refute all Hutton's pretensions to eternity (one of his sets of published letters was addressed to the Scottish savant). It was also the right order of magnitude to match the chronologists' calculations for the likely date of the Flood. It therefore supported his claim that "the present world" had started with a major physical event that could be equated with that biblical record. But Deluc was no biblical literalist. He conjectured that what had happened was a sudden interchange of continents and oceans: the antediluvial continents had collapsed below sea level, and the former ocean floors had been left high and dry as new postdiluvial continents. This was far from the biblical picture of a brief incursion of the oceans onto the land and their subsequent retreat. But it did explain the absence of human fossils, since any traces of the human world before the event would now be on the ocean floor. Conversely, it also explained the widespread marine fossils found on land, which in Deluc's view were relics of what he called "the former world."

So Deluc reconstructed the Earth's total history in terms of two contrasted "worlds" separated by a uniquely massive physical "*revolution*." As the title of his original *Letters* made clear, his goal was a history, and his focus was on establishing the historical reality of the physical event faintly recorded in the Flood narrative in Genesis. He recognized that its cause was a separate issue. While taking it for granted that the cause had been natural, he offered only brief suggestions that it might have been due to some kind of crustal collapse (Descartes' model continued, however indirectly, to be an inspiration). In contrast to his untiring attempts to date the start of the "modern world" as precisely as possible, Deluc left the timescale of the antecedent "former world" vague and unquantified. But he emphasized that its timescale must be immense by any human standard: he was no "young-Earth" literalist. Likewise his analysis of the Flood story itself was far from literal-minded and it took account of contemporary biblical scholarship. He claimed that this helped to clarify the religious meaning of the event, not to undermine its reality.

In his later writing, Deluc absorbed what other savants were discovering about the huge and varied pile of strata (to be described in the next chapter); on his travels he had seen some of the relevant evidence for himself. So he converted his undifferentiated "former world" into a sequence of phases in the antediluvian history of the Earth. Like Buffon he interpreted this history in terms of the traditional "days" of Creation, vastly extended as usual into periods of unimaginable duration. Unlike Buffon's "epochs," however, Deluc's sequence was no perfunctory parallel with Genesis but a conscientious concordance. But as with his analysis of the Flood narrative, the parallel did not need to be close, let alone literal. What mattered to Deluc was that the religious meaning of both narratives was not just conserved, but deepened, by these novel inputs from the world of nature. In contrast to Buffon, there was no programmed inevitability about Deluc's sequence of events and no pretension to predict the future. In contrast to Hutton, there was no reference to nature's intelligent design and no pretension to eternity. The causes of physical events were assumed to be natural throughout, yet those natural causes were set in a context of overarching divine "providence." Above all, Deluc's theory portrayed the Earth's own history as being just as contingent—and therefore unpredictable, even in retrospect—as the human history of the "present world" that was its culmination.

All this amounted to a decisive departure from most other theories of the Earth such as Buffon's and Hutton's. Deluc's theory was still ambitiously a Big Picture, but he had rejected a key feature of the earlier model: his was a scenario of the past and present but not of the future. He had rejected the unhistorical assumption that the future of the Earth was in principle predictable because it was fully determined by the laws of nature and, as it were, programmed in advance. The Earth in Deluc's theory was radically *contingent* and *historical*, without any less emphasis on the natural character of its causes. And the inspiration for this unmistakeably modern perspective could not be clearer. It came from Deluc's explicit and even fervent Christian theism.

Unlike Buffon and Hutton, the elderly Deluc lived on into the 19th century. But his theory of the Earth, like theirs and many others, was generally set aside as outdated. The whole genre of Big Pictures was felt to have outlived its usefulness. Nonetheless, specific elements from all three of these grand theories lived on, or were revived, and were used fruitfully in the hugely expanded exploration of the Earth's history that characterized the "*geology*" of the new century. The next chapter, however, stays within the later 18th century, in order to trace two related themes that have been lurking beneath the surface in this chapter. First, there was a dramatic, though usually implicit, enlargement of the timescale on which the events of the Earth's history were thought to have taken place. And second, historical ways of interpreting the Earth, similar to Deluc's, began to be developed and deployed in work that was less ambitious, but literally closer to the ground, than any all-explanatory Big Picture, Theory of the Earth, or scenario of past and future.

# 4

# Expanding Time and History

Few 17th-century savants had seen any good reason to doubt that the traditional timescale of world history, reinforced by the scholarly work of chronologists, was of the right order of magnitude. Those such as Woodward and Scheuchzer who were still active in the early 18th century saw no evidence in nature's own antiquities—notably fossils—to suggest that nature's history was any longer than the few millennia that had earlier been taken for granted. But during the second half of the 18th century evidence that did suggest a much longer timescale for the Earth began to accumulate rapidly. Far more important, however, was the way in which, increasingly, this expanded time was interpreted historically. Tracing what might have happened on the Earth in the course of time, when and how, proved much more significant than just trying to measure the total magnitude of the time involved. Time and history were expanded in tandem, but deep history turned out to matter far more than deep time.

The cultural climate of the Enlightenment, among its many other effects, fostered a greatly increased curiosity about the Earth and all its products, at every scale from huge physical features such as volcanoes and mountain ranges down to "specimens" or samples small enough to be collected, assembled, and displayed in museums. The 18th century, even more than the 17th, was a great age of collecting, particularly of objects relevant to the descriptive sciences of "natural history." Alongside specimens of animals and plants, relevant to zoology and botany, were those belonging to the third great "kingdom" of nature. These were the subject of the science of "mineralogy" (which at this time included the study of rocks and fossils as well as minerals in the narrower modern sense). Among all such specimens, fossils were of particular interest, precisely because they were now routinely regarded as relics of nature's own history. To call them nature's "coins," as Hooke had done, became a commonplace metaphor. And the range and variety of nature's coins—and therefore their potential value as evidence for the Earth's history—was greatly enlarged by the efforts of fossil collectors. Many kinds of fossils that in the 17th century had been regarded as highly problematic, and possibly not organic at all, came to be accepted as the remains of creatures that had once been truly alive. In most cases this was because collectors found specimens that were much better preserved than those previously available. For example, the distinctive and beautiful ammonites, which were often preserved only as molds or squashed flat on a surface of shale, proved to have been, beyond question, chambered shells coiled in an elegant plane spiral, somewhat like the highly prized shells of the living pearly nautilus. An even more striking example was that of the bullet-shaped solid objects known as "belemnites," which were often found in the same rocks as ammonites. The crystalline mineral structure of a belemnite had made it seem unlikely ever to have been a part of any organism. But during the 18th century much better specimens were found, which proved that the belemnite itself was just the most solid (and therefore most easily preserved) part of an otherwise delicate chambered shell rather like an uncoiled version of the nautilus. Belemnites could therefore be added to the ammonites, as fossils that could be treated as nature's own antiquities.

Woodward's great private collection of fossils, bequeathed at his death to the university in Cambridge, was just one of those which in the 18th century became accessible—at least to the socially respectable educated public—in museums in many cities around Europe. Some of these collections were also made mobile and still more widely accessible, when accurate trompe l'oeil engravings of the best specimens were published in lavishly illustrated volumes that were in effect "paper museums." Especially prominent in museums, both real and on paper, were the exceptionally well-preserved fossils that were found at a few specific localities scattered across Europe. Among them were quarries at Solnhofen in Bavaria, Oeningen near Konstanz, and Bolca near Verona (in modern terms they were all *Lagerstätten*, of which the Burgess Shale in British Columbia is now the most widely known). Each became famous, generating a profitable international trade in fine specimens that could be sold to the owners or curators of fossil collections. These spectacular fossils, preserving delicate structures in exquisite detail, were preserved in very thin layers of rock, which had obviously been deposited as very fine-grained muddy sediment, very

FIG. 4.1 An engraving of a fossil lobster on a slab of limestone from the famous locality of Solnhofen in Bavaria, as published in 1755 in a "paper museum" or lavishly illustrated set of volumes on natural history. Delicate fossils such as this, although squashed flat, were otherwise superbly preserved on the surface of thin layers of rock. They made it seem highly unlikely that this limestone, and similar rocks elsewhere, had been deposited during any brief Flood or Deluge, let alone in a violent event of any kind.

slowly and in very calm water. They undermined the earlier claim that all fossils could be attributed to a brief and violent "diluvial" event of the kind that Woodward, Scheuchzer, and many others had imagined. This was one of the many lines of reasoning that began to suggest that the Earth's timescale might need substantial stretching.

## STRATA AS NATURE'S ARCHIVES

These exceptional strata with superbly preserved fossils formed small parts of much thicker piles of rocks, which suggested even more strongly that the rocks as a whole might represent vast periods of time. However, this point could not be appreciated just by studying rocks and fossils indoors in a museum. It demanded fieldwork. In an age of self-conscious Enlightenment, there was a growing appreciation of the value of fieldwork, at least among naturalists if not the general public. Fieldwork was by definition done outdoors: literally in fields, or in quarries and along coastal cliffs, or high on mountains and deep in mines. It was often strenuous work, far from the comforts of civilization. It could not be delegated to underlings but needed to be done by the naturalist himself (it was almost always "him," because outdoor fieldwork by women, unlike the indoor work of assembling collections of specimens, was severely restricted by the social conventions of the time). The naturalist had to see with his own eyes what the rocks or the mountains looked like "in the field," often relying on the local knowledge of people of lower social class such as peasants, quarrymen, and miners to guide him to the most significant spots.

Many naturalists did fieldwork not primarily to satisfy their scientific curiosity but for much more practical reasons. It was a sign of the times that in the later 18th century several European governments founded mining academies to train scientific personnel for their burgeoning mineral industries (Britain was an exception, leaving mining entirely to private enterprise). What was needed for finding and exploiting fresh mineral resources in any specific region was to work out and describe the underground structure of the rocks, which could then guide the opening of new quarries and the sinking of shafts for new mines. Detailed three-dimensional surveys of this kind generated a new branch of mineralogy, which was called the science of "geognosy" ("Earth-knowledge"). Those who called themselves "geognosts"

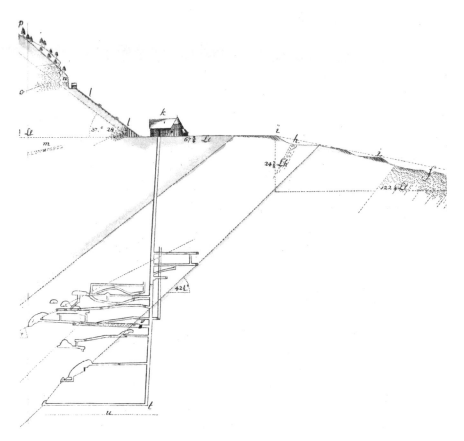

FIG. 4.2 A section through a mine in the hills of the Harz region in northern Germany, showing a vertical "*shaft*" and horizontal "*adits*" leading to the open spaces from which a mineral deposit is being extracted. These underground workings, combined with the rocks visible at the surface, enabled the *structure*—in this case a set of rocks inclined at about 45°—to be mapped in three dimensions: this was the primary task of the science of geognosy. This section was published in 1785 by Friedrich von Trebra, a geognost who was one of the first graduates of Saxony's great mining academy at Freiberg.

were primarily concerned to describe the composition and structure of the Earth's crust, not to propose causal explanations of what they observed, let alone to reconstruct the Earth's past history. They often contrasted their own soberly factual investigations with the kind of theorizing that Deluc proposed to call "geology," which they tended to dismiss as no better than fanciful speculation.

Yet some basic ways of interpreting rocks and their structures in terms of causes and history seemed reliable and almost unavoidable, even to the most hard-nosed geognosts. They found that in many regions the rocks fell into two major categories, which recalled those that Steno had recognized in Tuscany a century earlier. The rocks that were lowest in position and therefore apparently the oldest, such as granites, schists, and slates, were called "*Primary*" (or "*Primitive*") rocks, and were often attributed to the earliest phase of the planet's history. Since no fossils were found in them they were usually assumed to date from before the origin of life. Overlying them were varied rocks such as sandstones, shales, and limestones. These were obviously younger, and some of them seemed to have been formed from the debris of Primary rocks, so they were called "*Secondary*." Many contained fossils, sometimes in abundance. Most of these rocks, both Primary and Secondary, showed layers or strata and were said to be "*stratified*," which implied that they had accumulated gradually, layer by layer (a few rocks, such as granite, were massive or "*unstratified*," which gave some support to Hutton's claim that they had originated in a quite different way). Rocks of all kinds could also be grouped into distinctive units that were known as "*formations*," the surface outcrops of which could often be traced across country and could be depicted on maps. Among the Secondary rocks, one well-known example was the Chalk formation, which was widespread in northwest Europe; its most prominent component rock was the brilliant white limestone seen in the white cliffs on both sides of the narrow Straits of Dover that keep England and France apart. Another was the Coal formation, which included, along with many intervening strata of sandstone and shale, the thin but valuable seams of coal that were literally fueling the nascent Industrial Revolution, most notably in Britain.

Geognosts tried to classify all these varied rock formations, just as botanists classified plants and zoologists classified animals. For example, Abraham Werner, a leading geognost who taught at Saxony's mining academy in Freiberg, published a *Brief Classification and Description of the Different Species of Formations* (*Kurze Klassifikation und Beschreibung der verschiedenen Gebirgsarten*, 1787). This was intended to bring some kind of order to what he himself had seen in his extensive fieldwork, and what he knew his contemporaries were finding in various other regions (Werner's booklet, like Steno's a century earlier, was intended as a trailer for a larger work that never appeared). As its title made plain, it was a classification and description of *species* or kinds of things, just like those devised by botanists and zoologists for plant and animal species. It was not, or not primarily, an interpretation of formations in terms of the causal processes by which they had been formed, let alone in terms of the Earth's history. Within each category, Werner's listing of the various "species" did not indicate any invariable sequence in which they had accumulated.

More broadly, however, geognosts, like Steno, did take the distinction between Primary and Secondary formations to reflect the outline of a directional kind of history for the Earth: from a period before the first appearance of life to a subsequent period with plenty of living creatures. And Werner later proposed a category of "*Transition*" formations, for those that were intermediate not only in their position in the pile, but also in that they were found to contain just a few rather obscure fossils. At the most recent end of this outline of the Earth's history, he and other geognosts had already recognized a category of "*Alluvial*" ("washed out") deposits of loose sand and gravel. These were found overlying and therefore obviously more recent than any of the solid rocks; and they were composed, equally clearly, of debris derived from those older rocks, such as pebbles or boulders of distinctive granites and limestones that could be matched in the bedrock elsewhere.

In several parts of Europe, the careful fieldwork of geognosts revealed piles of Primary and Secondary rock formations (and overlying Alluvial deposits). They were usually more or less concealed beneath a cover of soil and vegetation, but were visible here and there in cliffs and river beds, quarries and mines. What was unexpected was the discovery of their huge total thickness and sheer diversity. The implication of this for understanding the Earth's history was almost inescapeable. It became highly implausible to suppose that all the Secondary formations had been formed during any single brief Flood event, whether it was conceived as a tranquil rise and fall in global sea level or as a violent transient mega-tsunami. After the middle of the 18th century, any connection between the Secondary formations and the biblical record of the Flood was rarely asserted. In effect, the kind of diluvial theory represented by Woodward's and Scheuchzer's global Deluge or mega-Flood event faded from debates among savants (it lingered on among the less well-informed public). The Flood itself was far from being eliminated from their discussions, but those who argued for its historical reality no longer claimed that it could be held responsible for all the Secondary formations and their fossils. From now on, the Secondary formations were attributed to the *antediluvial* history of the Earth. In contrast, the possible traces of the subsequent Flood were confined to the Alluvial deposits: although modest in thickness they were widespread enough to be plausible relics of an allegedly global event of some kind. This new geognostic understanding of the structure and sequence of the formations forming the Earth's crust clearly implied that even the Flood—if in reality there had been such an event—must have been relatively recent in the Earth's history, although it was such a distant event in human history. This again suggested that the total history of the Earth might be far lengthier than the few millennia that the chronologists had calculated.

Furthermore, even the most comprehensive collections of fossils contained no specimens of human fossil bones or artifacts, or none that was not dubious or highly controversial. Scheuchzer's earlier claim to have identified "a man a witness of the Flood" has already been mentioned. There were several similar claims later in the 18th

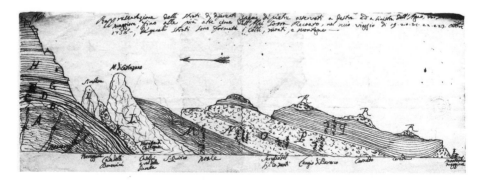

**FIG. 4.3** A section or slice through the Earth's crust, drawn by the Venetian geognost Giovanni Arduino in 1758, showing the huge pile of varied rocks visible on the sides of a valley running from the north Italian plain up into the Alps (the arrow points to the north). The lower and therefore older rocks (left) were *"Primary"*; the upper and younger (right) were *"Secondary."* Those labeled M through Q are marked with thicknesses totaling many thousands of feet (in modern terms they range in age from Permian to Oligocene). Such thick piles of rocks made it seem likely that the strata represented correspondingly vast spans of time, far beyond the few millennia of the traditional timescale. Although this drawing was never published, it was almost certainly shown—and perhaps even demonstrated on the spot, *in the field*—to the naturalists who visited Arduino in the course of their travels. The length of the section is about 30 km; the vertical scale is greatly exaggerated to make the structure clear.

century. One naturalist reported that an iron key had been found in solid Secondary rock in a quarry just outside Paris, but the object itself had not been preserved and there was only a quarryman's word for it. Another described a finely polished stone axe that had been found in a quarry outside Brussels, but again its significance depended on the reliability of the quarrymen's testimony that it had been found within the solid Secondary rock and not loose on the surface. Yet another naturalist reported finding a couple of human bones among abundant fossil animal bones in a deposit inside a German cave, but they might have come from some much more recent burial dug into the floor of the cave. Such claims were all too uncertain to be relied on, or too dependent on the word of people of lower social class, who might be reporting what they knew the naturalist would reward them for finding. Although of course all this was negative evidence, the continuing failure to discover any unambiguous traces of *human* life among the abundant fossils of the Secondary formations, or even the Alluvium, did suggest that these deposits must have accumulated during vast periods of pre-human time, and therefore that human beings must have been very recent newcomers on Earth (this was the widespread hunch that Buffon made explicit).

## VOLCANOES AS NATURE'S MONUMENTS

Of all the natural features that called for fieldwork, it was volcanoes that most caught the imagination of educated people in the 18th century. The active volcano Vesuvius, looming over the great city of Naples in the south of Italy, was for Europe's savants (and aristocratic tourists) an unmissable item on their Grand Tour of the sites and sights of the ancient Classical world. To see Vesuvius—and even perhaps to climb to its summit crater, if the volcano was in a quiet phase—was considered almost as important as it was to see the ruins of Herculaneum and Pompeii, the buried Roman cities at its foot. These had been discovered earlier in the 18th century and were being excavated with sensational results. Their destruction in Roman times, in the well-recorded eruption of AD 79, represented a striking conjunction of human history with the history of nature. Sir William Hamilton, the British ambassador in Naples (and later the husband of the glamorous Emma, Admiral Lord Nelson's lover in a famous *ménage à trois*),

**FIG. 4.4** The 1767 eruption of Vesuvius, with a lava flow threatening the city of Naples at its foot. This etching was published in Sir William Hamilton's *Campi Phlegraei* (1776; its text was not in Italian but in French and English), his lavishly illustrated description of the volcanic region. The huge cone of Vesuvius itself—here brightly lit by the eruption—had clearly been built up from successive lava flows and falls of volcanic ash, the more recent of which could be dated from historical records going back to the notorious eruption of AD 79. The dark hill to the left, Monte Somma, was interpreted as a relic of an even older cone, which appeared to take the sequence of eruptions still further back in time. This scene was drawn in the dramatic style characteristic of the landscape art of the time, which was also well suited to the depiction of features of scientific importance such as this.

made himself a leading expert on both volcanoes and antiquities. Visitors to Naples might take home with them not only antiquities such as ancient "Etruscan" vases, which Hamilton the antiquary showed were in fact Greek, but also volcanic rocks and minerals, samples of which Hamilton the naturalist sent to the Royal Society in London to illustrate his reports on specific eruptions. Hamilton recognized that the many historically recorded eruptions of Vesuvius since AD 79, and even more clearly those of the much larger volcano Etna on the island of Sicily, must have been preceded by a lengthy sequence of similar eruptions, stretching back into the unrecorded times before even the ancient Greeks arrived. And these huge cones of volcanic ash and lava flows were apparently built on foundations of still older rocks. All this suggested, at the very least, that those unrecorded times might have been far longer than had previously been suspected.

Such reasoning was reinforced by the sensational discovery of extinct volcanoes, hundreds of miles from any active ones, in France's hilly Massif Central. Reports that volcanic rocks had been seen in this remote region were followed up by Nicolas Desmarest, a naturalist who was traveling there primarily to report to the French government on local industries. Desmarest had earlier accompanied a young aristocrat on a Grand Tour that included a visit to Naples and Vesuvius, so he was well prepared to recognize volcanic features, even under a concealing cover of more recent vegetation. In the province of Auvergne he found many unmistakeable cones of loose volcanic ash topped with craters, with equally unmistakeable flows of solidified lava emerging from them and extending for miles down the valleys. Yet there was no historical record of any volcanic eruptions anywhere in France, nor any hint in local folklore that any had ever been witnessed. Desma-

CRATERE DE LA MONTAGNE DE LA COUPE, AU COLET D'AISA,
Avec un Courant de Lave qui descend inutuusseure à un pavé de basalte prismatique.

**FIG. 4.5** An extinct volcano in a remote part of France's hilly Massif Central, with a narrow stream of lava flowing—somewhat diagrammatically—from its summit crater. At the river's edge in the foreground, erosion of the lava has revealed that it is composed of the rock called *basalt*, with its distinctive vertical columns. This engraving of the Coupe d'Aizac in the province of Vivarais (south of Auvergne) was published in 1778 by Desmarest's younger contemporary Barthélemy Faujas de Saint-Fond, who later became, in Paris, the world's first professor of "geology."

rest found that one small lake, on which a late-Roman poet (and early Christian bishop) had enjoyed the fishing, owed its very existence to a lava flow that had blocked the valley and formed a natural dam. This was strong evidence that all such eruptions must have been further back in time than the Roman period, perhaps even before the earliest human records anywhere. The extinct volcanoes of central France, which soon became well known in scientific circles, suggested that volcanic activity might stretch far back in the Earth's history.

This possibility was made much more likely as a result of Desmarest's detailed study of the whole volcanic region in Auvergne; his maps were made for him by a military surveyor who had been made redundant when the Seven Years War ended. He discovered that the lava flows on the floors of the valleys were matched by similar rock capping some of the adjacent hills. If this too was volcanic in origin, it must have flowed as lava down valleys that had long since disappeared. Desmarest argued that the hills that had channeled these ancient flows must have been worn down by the slow but observable action of rain and rivers—one of Deluc's "present causes"—eventually leaving the ancient lava high and dry on the hilltops above the newer valleys (the relatively soft bedrock in this region would have been eroded much more rapidly than the hard lava). In effect, hills and valleys must have exchanged places. If the process of erosion was indeed as slow and steady as it seemed, this hinted that the time needed to produce the present landscape of Auvergne must have been immensely extended, far beyond the total span of human history, and it did not seem to have been interrupted by any episode like the Flood.

Significantly, Desmarest borrowed from the chronologists their key concept of "*epochs*," transposing it from human history into nature's history. And he borrowed from Steno—whose work was still well known in scientific circles—the method of penetrating back from the observable present to the more obscure past, before reversing direction and reconstructing a true history from past to present. In his early reports he referred to the most recent set of volcanic cones and lava flows as those of a "first epoch," since it was nearest to the present; but later he renamed it a "third and last epoch," reflecting its place in the history of the region. (Desmarest was using the language of "epochs" long before Buffon—a much more powerful figure in the scientific world—upstaged him by publishing *Nature's Epochs*.) And he referred

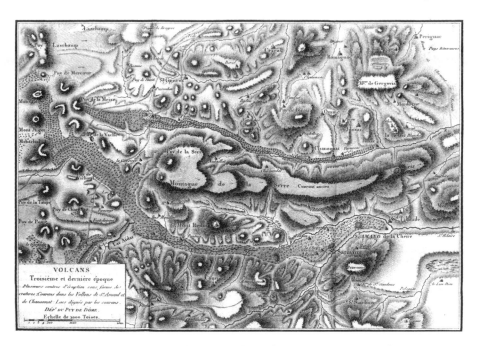

FIG. 4.6 Desmarest's detailed map of a small part of Auvergne, showing two lava flows
(stippled) emerging from small volcanic cones to the west (left) and flowing eastwards down
two parallel valleys. Between them is a long narrow plateau, also sloping to the east and capped
by hard basalt, which Desmarest interpreted as a far older lava that had once flowed down a
similar valley. Some other nearby hills, also capped by basalt, were interpreted as relics of other
ancient lavas, even more severely eroded. Desmarest identified all these lavas as the products of
two distinct "epochs" of volcanic activity in this area (a third and still more ancient epoch was indicated by the pebbles of basalt found in the underlying bedrock). One of the most recent flows
has ponded back a small lake (Lac Aidat, lower left), which was known to have been in existence
in Roman times; so even this "third and last epoch" dated from before recorded human history.
This map was not published until 1806, but it was based on one that Desmarest displayed in 1775
at the Académie des Sciences in Paris; it depicts an area about 20 km across.

specifically to the newly excavated ruins of Herculaneum (much better preserved than those of Pompeii) as directly analogous to his own
reconstruction of the past history of Auvergne.

However, Desmarest's conclusions depended on his claim that the
rock on the hilltops in Auvergne, known as "*basalt*," was genuinely
an ancient lava. In fact the origin of this dark fine-grained rock was
highly controversial (the famous "black basalt" stoneware invented by
the industrial potter Josiah Wedgwood imitated the rock's appearance
quite accurately). Basalt was often jointed into strikingly regular hex-

agonal columns: examples that became famous throughout Europe in the 18th century were the "Giant's Causeway" on the northern Irish coast, and "Fingal's Cave" on the Isle of Staffa off the west coast of Scotland. And thick layers of basalt were often found in the middle of piles of Secondary rocks such as sandstone, shale, and limestone, all of which had apparently been deposited in vanished seas: a famous example, complete with vertical columns, was visible in the cliffs of Salisbury Crags, rising high above Hutton's home in Edinburgh.

Yet Desmarest claimed that basalt was definitely a lava. For example, he found that hexagonal columns could also be seen in some of the relatively recent and unquestionably volcanic lavas in Auvergne. Such field evidence eventually convinced most of the relevant naturalists that basalt must be volcanic, though an influential minority, Werner among them, continued to argue that it was some kind of hardened sediment (techniques for studying the micro-structure of fine-grained rocks, which would have helped resolve the puzzle, were not developed until the mid-19th century). Since all these naturalists were well educated in the Classics, the controversy was described jokingly as pitting "*Vulcanists*" against "*Neptunists*," the devotees of one ancient god against those of another. Eventually most of them agreed that basalt belonged in Vulcan's realm, while Neptune continued his sway over most other rocks. One prominent naturalist dismissed the whole basalt controversy, with some justification, as a mere storm in a tea-cup, because it involved the classification of just one kind of rock among many. It would indeed have been a relatively minor scientific squabble, had it not had one important implication for the Earth's own history. The recognition of basalt as a volcanic rock showed that volcanic activity much like that of "the present world" extended right back into the times of "the former world" when the huge piles of varied Secondary rocks had accumulated. Volcanoes were evidently an integral part of the Earth's economy, as it were, and not just a superficial feature of its present state (perhaps fueled—it had been suggested—by the underground combustion of coal seams among the Secondary rocks).

## NATURAL HISTORY AND THE HISTORY OF NATURE

Desmarest referred to the antiquaries' excavation of Herculaneum as analogous to what he was doing with the extinct volcanoes of Auvergne. This potent analogy with human history was deployed even more extensively by a younger naturalist who emulated Desmarest's work in another part of the Massif Central, the province of Vivarais. As a young man, Jean-Louis Giraud-Soulavie served as parish priest in a village that happened to be in full view of one of the extinct volcanoes, so he knew at first hand the intriguing questions they raised. After moving to Paris to make a career as a savant, he described his extensive fieldwork and set out his interpretation of it in a seven-volume *Natural History of Southern France* (*Histoire Naturelle de la France Méridionale*, 1780–84). In fact this was far more than a conventionally descriptive "natural history." It was shot through with what Soulavie treated as the still novel and under-exploited idea of reconstructing nature's own *history* (in the modern sense). He called himself "nature's archivist" and claimed to be compiling the "annals of the physical world" by working out the "physical chronology" of the volcanoes. The various rock formations, including the basalts that he interpreted as ancient lavas, were nature's "monuments" and "inscriptions," which recorded a long sequence of nature's "epochs" in that region.

Like Steno and Hooke long before, but far more thoroughly, Desmarest and Soulavie were deliberately transposing the methods and concepts of the chronologists and the antiquaries from the human into the natural world, from the brief span of human history into the almost unimaginable depths of the Earth's own history. It is no coincidence that they did so at a time not only of exciting new discoveries in archaeology, but also of outstanding scholarship in the writing of human history, as for example in Edward Gibbon's celebrated *Decline and Fall of the Roman Empire* (1776–88). And it is unsurprising that Soulavie later turned to Gibbon's kind of history, when he published a detailed study of French politics under the *ancien régime*. His and Desmarest's persuasive demonstrations of how *nature's* history could be reconstructed in a similarly detailed and reliable way, from evidence equally carefully observed and checked, was not immediately followed up by many other naturalists: as Soulavie pointed out, this kind of reasoning was still novel and unfamiliar. But in the longer run their use of an analogy with the writing of *human* history became the decisive strategy for reconstructing the history of the Earth.

In addition to all the volcanic rocks in Vivarais, Soulavie described a pile of three Secondary formations. He found he could recognize

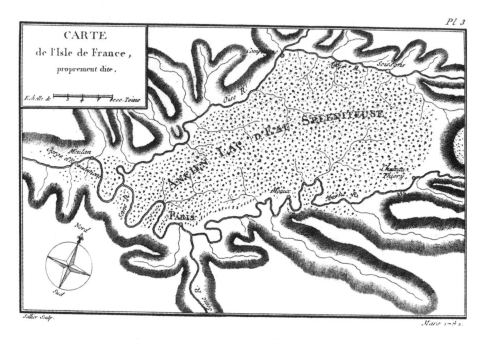

FIG. 4.7  A vast former lake (stippled) to the north of Paris, reconstructed from the evidence of a localized deposit of gypsum within the pile of Secondary rocks in this region. This map (in modern terms "*palaeo-geographical*") was published in 1782 by the French naturalist Robert de Lamanon; he interpreted the gypsum or selenite as (in modern terms again) an evaporite deposit from a "former lake of selenitic water"; it is shown as having been about 120 km in length. This is an example—still quite rare in the later 18th century—of an explicitly *historical* interpretation of features derived from a structural or geognostic survey of rocks and minerals.

each across the region by its distinctive set of fossils (in modern terms the formations were Jurassic, Cretaceous, and Miocene in age). Other naturalists who had studied Secondary rocks in other parts of Europe had also noticed this kind of relationship between formations and their fossils. It had not yet been given detailed attention, but it was clear, for example, that the lower and older Secondary formations (mostly, in modern terms, Mesozoic in age) often contained ammonites and belemnites, whereas the upper and younger ones (now termed Cenozoic) never did; conversely, the fossil shells in the younger formations were more like the shellfish living in present seas than those in the older. However, it was far from clear how all this should be interpreted. Soulavie claimed that the sequence of fossils in his formations recorded a part of the true history of life. But other naturalists thought it more likely that the differences in the fossils simply reflected the changing environmental conditions in which the animals had lived and the sediments had accumulated (in modern terms, their *facies*). The older formations might have been deposited in very deep water, preserving the remains of shellfish that still live in that environment.

This was far from being implausible. As Hooke had realized long before, all too little was known about the world's plants and animals: every long-distance voyage or expedition brought back to Europe spec-

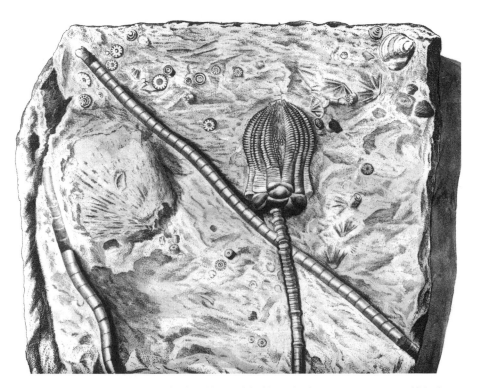

FIG. 4.8 An "encrinite" or fossil sea-lily on a slab of Secondary limestone: an engraving published in 1755 in a "paper museum" of natural history. These well-known fossils—highly prized by fossil collectors—might have been assumed to be extinct; but similar sea-lilies were dredged up from deep water, and were first described in print in 1755 (by coincidence, in the same year). Such "living fossils" made the very idea of extinction highly uncertain, and any reconstruction of the *history* of life equally problematic. The small circular objects also shown on this slab of rock were among the many fossils that had earlier been considered dubiously organic in origin; but well-preserved specimens, such as this, showed that they were separated segments of the flexible stalk of a sea-lily. Although superficially plant-like, sea-lilies or "crinoids" were soon recognized as basically similar to starfish, brittle-stars, and sea-urchins; later they were all classed (as they still are) as "echinoderms."

imens of plants and animals previously unknown. The ocean depths were even more obscure. It seemed quite possible that ammonites, for example, might still be flourishing there. The discovery of what are now called "*living fossils*" seemed good indirect evidence for this. One of the most striking cases was the living sea-lily (in modern terms a "*crinoid*") that a long plumb-line happened to bring up from deep water in the Caribbean. This specimen was obviously similar, though not identical, to the fossil sea-lilies that were already well known in some of the Secondary formations. The discovery of a sea-lily as a "living fossil" in the ocean depths made it seem quite likely that ammonites too might in due course be found alive there (the modern discovery of the coelacanth, a "living fossil" fish now known to be quite common in deep water off the Comoro Islands in the Indian Ocean, is a salutary reminder of the continuing validity of this point.)

Since many or all of the commonest fossils might still be flourishing somewhere as "living fossils," the formations in Vivarais or anywhere else could not yield any unambiguous global history of life, although they were clearly the record of a sequence of *local* changes in the Earth's physical state. The Earth might have supported much the same kinds of plants and animals throughout its history, even if their geographical distribution had changed in the course of time. This meant, for example, that the conjectural steady-state "system" of the Earth proposed by Hutton (and earlier by Buffon, before he adopted the model of a cooling Earth) might be more on the right lines than Deluc's directional and strongly historical system. Deluc's concept of a sequence of distinctive periods—inspired by the Genesis narrative, though not following it closely—would become more plausible than a steady-state system, only if his "former world" could be shown to be quite distinct from the "present world," everywhere on Earth, and in its animals and plants as much as in its physical character. The remains of shellfish and other marine animals, which were (and still are) by far the commonest fossils, were ambiguous in this respect, owing to the strong possibility that many or all of them might still be in existence as "living fossils." But living land animals, or at least the more conspicuous ones such as large mammals, were much better known. So they might be better able to act as a benchmark for comparisons with similar animals of the "former world."

This is why, in the late 18th century, the huge fossil bones and teeth that were often found in Alluvial deposits became a focus of naturalists' attention. Those found in Europe had long been attributed by the uneducated to antediluvian giants, but early anatomists had shown that they were certainly not human. Many were identified instead as the remains of elephants, and were then attributed to those that Hannibal had famously imported from north Africa for use in his wars against the Romans. However, many new discoveries of similar bones all around Europe—but also in Siberia far to the east, where indigenous people called them "*mammoth*," and in North America far to the west—switched attention to possible natural causes. For example the Flood, if in reality it had been a huge tsunami, might have swept the carcasses of elephants out of their tropical habitats in Africa and Asia into more northerly regions (this explanation did not work so well in the North American case).

The puzzle was deepened, however, by evidence that some of these bones and teeth belonged to an animal unlike any species known alive. Elephant-like tusks and hippopotamus-like teeth seemed to have belonged together in the same mammal. This "Ohio animal," so called from a famous locality in the wild west of the British colonies in North America, had evidently been widespread in northerly latitudes in both the Old World and the New. Some naturalists took it to be decisive evidence of genuine extinction: Buffon thought it might have been adapted to even hotter conditions than the present tropics, and then have been killed off as the Earth cooled down. But others, such as Thomas Jefferson, later suspected—and probably hoped, for reasons of national pride—that it was still alive and well and living in the poorly explored interior of what by then had become the independent United States: as its president, he instructed Lewis and Clark, on their famous overland expedition to the west coast, to look out for it. In view of such uncertainties, it would take more than this one problematic case to establish with any confidence that the complete extinction of any species—a possibility that many naturalists found hard to swallow—was a regular feature of the natural world. More generally, it would be hard to establish that a true *history* of life could be based on all the fossils found in the sequence of rock formations and the still more recent Alluvial deposits.

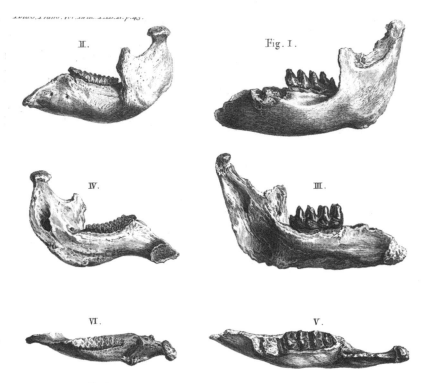

**FIG. 4.9** The lower jaw of the enigmatic fossil "Ohio animal" (right), later named the "*mastodon*," compared with that of the living elephant (left), each shown in external, internal, and dorsal views. These engravings illustrated a paper read at the Royal Society in London in 1768 by the surgeon and anatomist William Hunter. He argued that "the American *incognitum*" was a distinct species unknown alive, and that this was probably a true case of extinction. Other naturalists, however, thought it more likely that the animal was still alive as a "living fossil" in some unexplored part of the world.

## GUESSING THE EARTH'S TIMESCALE

As already pointed out, most of these hints or suggestions of a vast timescale for the Earth could not be fully appreciated just from studying fossils or rock specimens in a museum, still less from merely reading books in a library. The possibility of a hugely extended history of the Earth, almost all of it probably pre-human, was most convincing to those naturalists who had seen for themselves, *in the field*, the sheer scale of the piles of rock formations and the size of the great volcanoes. Their growing suspicion that vast spans of time must be involved generally remained both implicit and unquantified. This was not for fear of criticism from church authorities, but for the much stronger reasons that they had no reliable way to measure the time involved, and that they had no wish to be thought merely speculative. Yet their unpublished informal remarks (where any have survived in the historical record) show that by the later 18th century many of them were thinking—openly, routinely, and almost casually—in terms of at least hundreds of thousands of years, or even millions, for the accumulation of the piles of strata and of the still more recent volcanoes. Werner, for example, is said to have talked of perhaps a million years for the huge rock pile with which he was familiar, and similar guesses were made by others too. Such an amount of time may seem pitifully inadequate to modern geologists, but it does show that their predecessors in the later 18th century had already taken the crucial *imaginative* step of thinking of the Earth's own history in terms that vastly exceeded the traditional few millennia. At the time, imagining even hundreds of thousands of years had just as great an impact as imagining billions would have had. Even the lower order of magnitude was more than enough to dwarf the whole of known human history. It was, in a literal sense, an almost *inconceivably* vast span of time.

By the later 18th century, then, the field evidence was convincing enough for many naturalists to assume some kind of extremely long timescale as a matter of course in all their reasoning about the Earth. Buffon, who explicitly proposed and published such a timescale, was criticized not for its order of magnitude—which if anything was on the low side, by the standards of his contemporaries—but for the dubious conjectures that underlay his startling precision about it. Hutton, who took for granted a timescale without limits, was criticized not for its undefined magnitude but for its scarcely veiled eternity. Deluc was more typical, in that he declined to put figures to any period except the most recent (the "present world"), and assumed a vast but unquantifiable magnitude for the rest (the "former world"). Never in the subsequent history of this kind of science did those with the relevant field experience doubt that the Earth's timescale must dwarf the totality of recorded human history; in contrast, the opinions of the general public, who lacked this first-hand knowledge, often remained quite different. By the later 18th century, savants took it as much for granted that the Earth was almost *inconceivably* ancient as their predecessors a century earlier had assumed that its history could be measured in mere millennia. From now on, any savant who proposed or inferred a very long timescale, or just took an extremely ancient Earth for granted, was pushing at an open door (it is a modern misconception that this crucial change of perspective had to wait for the geology of the early 19th century, or even for Darwin's evolutionary theory still later in that century).

The many savants who regarded themselves as Christian believers found this extended timescale no more problematic or disturbing than their unreligious contemporaries. As mentioned earlier, biblical scholars had recognized the ambiguity of the key word "day" in the Genesis narrative, long before any evidence derived from the natural world began to throw doubt on the traditional and commonsense assumption that it denoted an ordinary day of twenty-four hours. If the "days" referred instead to something like the future "day of the Lord," or to major turning-points within the divine drama of Creation, world history might be of much greater length than a few millennia, without affecting the authority of the biblical text or—much more importantly—its religious meaning. So it is not surprising that those who in the later 18th century began to take a very long timescale for granted were not criticized by church authorities, except occasionally and in very specific local circumstances (contrary to modern myths of universal conflict, repression, and persecution). What mattered far more, from a religious point of view, was that the finite "createdness" of the whole universe should continue to be affirmed, against those who claimed it was eternal and therefore uncreated. This was of course a philosophical and theological matter that could not be settled by reference to scientific observations. On such a fundamental issue, it was the religious, not the sceptics, who often felt themselves to be an embattled minority. Deluc, for example, far from representing a dominant viewpoint or a domineering orthodoxy, felt himself to be defending Christian theism against a culturally more powerful majority of Enlightenment deists and atheists, the "cultured despisers" of religion (as Friedrich Schleiermacher famously called them).

It was therefore quite easy for Christian savants such as Deluc to assimilate and endorse a previously unimaginable timescale for the history for the Earth. They were well aware that at just this time the *historical* methods of textual interpretation already in use for under-

standing Classical literature were beginning to be applied consistently
to biblical texts (it is another modern misconception that biblical criticism did not get going until well into the 19th century). The striking
development of biblical criticism during the 18th century—far beyond
that of Ussher's time—was not the inevitably anti-religious weapon
that it has often been portrayed as being. Biblical criticism was a two-
edged sword (and of course, as pointed out already, it was "criticism"
in the same sense as literary, musical, or artistic criticism). Often it
was indeed intended to undermine or invalidate traditional religious
beliefs, usually in the service of secularist political goals. But it could
equally well be motivated by the hope of reaching a deeper understanding of the *meaning* of the texts for their original writers and
readers, as a necessary precondition for translating that meaning into
terms that were relevant and of value to the contemporary practice
of religious faith. The Creation narrative therefore continued to be a
fruitful source of inspiration, for thinking about the Earth as a whole,
long after the story had ceased to be interpreted—by anyone familiar
with the new scientific evidence—in the earlier "literal" manner.

In conclusion, what was far more significant than the stretching
of the bare magnitude of the Earth's timescale was the character of
the Earth's history that was reconstructed within this greatly extended
span of time. As already suggested, deep history mattered far more
than deep time. More specifically, the biblical narrative of Creation
*pre-adapted* savants—unless they were actively averse to anything
religious—to find it easy and congenial to think about rocks and fossils, mountains and volcanoes, as evidence for the Earth's *history*. The
Genesis story of an intelligible but contingent sequence of unrepeated
events, by which the Earth and its living things had been brought into
their present state, could readily be expanded from a notional "week"
into an incalculable span of time; the first five "days" of Creation before human beings arrived on stage could readily be enlarged from a
mere scene-setting prelude to become by far the longest part of the
entire drama. This is just what happened, at least among savants, in
the later 18th century.

Yet at the end of the century the details of this drama remained
quite obscure. It was uncertain whether or how far the Earth in the
deep past had been significantly different from its present state. In
particular, it was far from certain whether *life* had had a true history,

or whether the same kinds of plants and animals, more or less, had always been around. It was clear that the history of the Earth had been almost unimaginably lengthy in comparison with the whole of human history; but it remained unclear whether human beings could ever *know* what had happened in this deep pre-human past, in detail and with the kind of confidence that could now be given to human history. This was the fundamental problem that remained to be tackled in the early 19th century, and it is the topic of the next chapter.

# 5

# Bursting the Limits of Time

## THE REALITY OF EXTINCTION

By the late 18th century, naturalists with first-hand experience of the relevant evidence had reached an implicit consensus that the total timescale of the Earth's history must have vastly exceeded the few millennia that earlier generations had taken for granted as common sense. Yet what exactly had happened during this newly extended span of deep time remained profoundly uncertain. The piles of rock formations that were explored and described in several parts of Europe suggested that an initial period before any living things existed (represented by the Primary rocks) had been followed by another in which the seas had teemed with life (the Secondary rocks, often with abundant fossils), and that human beings had appeared on the scene only at the last moment (being apparently unrepresented by any fossils at all). Yet even this bare outline of the Earth's deep history was uncertain and controversial. Buffon, for example, had expanded it into a secular version of the Creation narrative in Genesis, but his account was based on thin evidence and could easily be dismissed as speculative science fiction. Deluc had argued, from much more extensive evidence, that there had been a uniquely disruptive "revolution" in the relatively recent past; but since he identified it with the narrative of Noah's Flood it was forcefully denied by those, such as Hutton, who dismissed any such input from a religious source and any such exceptional deviation from the normal course of nature. In effect, Hutton rejected both Buffon's sequence of distinctive "epochs" and Deluc's dramatic recent "revolution"; he claimed instead that the Earth was a smoothly operating natural "machine," continuing its repetitive cycle of similar "worlds" from and to eternity.

Even if all such Big Pictures were set aside as over-ambitious and premature, the researches of other naturalists—more modest in scale but often more thorough in their underlying fieldwork—had only limited success in providing a more reliable reconstruction of the Earth's deep history. In particular, it remained highly uncertain whether life itself had had any true history, apart from the very late arrival of human beings. The discovery of "living fossils" in the ocean depths suggested that any apparent sequence in the earlier history of life might be an illusion based on inadequate knowledge of the present world. For example, the abundance of ammonites and belemnites in the lower and older Secondary rocks, and their absence from the younger formations, might be due simply to their deep-sea habitat, then and now. They might not be extinct at all. Fossils could not be treated as reliable "coins" or "monuments" of nature, giving reliable evidence about earlier phases in the Earth's deep history, unless extinction had in fact been a regular feature of the natural world. But this was just what was utterly uncertain: the only well-documented cases of extinction, such as the famous flightless dodo on the Indian Ocean island of Mauritius, were those due to recent *human* agency. And such doubts about the reality of extinction in nature were strongly reinforced by the gut feeling, as it were, that neither the caring providential God of Judeo-Christian theism, nor the almost impersonal Supreme Being of Enlightenment deism, would or could have allowed any created species to go extinct, the exceptions being due to the actions of sinful or at least careless human beings.

The question of extinction was therefore central to any attempt to understand the deep past. This is why fossil bones, in particular, became such a focus of attention among naturalists in the years around 1800. The fossil "Ohio animal" was not decisive on this point—even if it really was distinct from any known living mammal—since it might

still be alive as a "living fossil" in the poorly explored interior of North
America or central Asia. But if other fossil bones proved to be similarly distinct from those of any living species, the case for genuine
extinction would become stronger. The best hope of resolving this uncertainty was therefore to compare fossil bones with those of living
animals, in greater detail and over a wider range of species.

There was one naturalist who was in the right place at the right
time to do this, and who also turned out to have just the right skills in
exceptional measure. Shortly after the Terror, the bloodiest phase of
the Revolution in France, the young Georges Cuvier was appointed
to a junior position at the Muséum d'Histoire Naturelle, which had
just been "democratized" from the Parisian institution that Buffon
had ruled autocratically under the royalist Old Regime. At the Muséum Cuvier had access to the world's finest collection of zoological
specimens; it was the best possible database for comparing fossils with

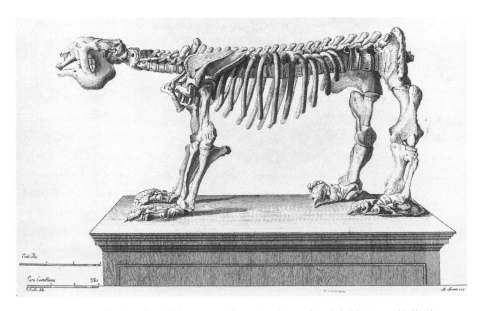

FIG. 5.1 The huge fossil skeleton—nearly four meters long and two in height—assembled by the
Spanish naturalist Juan-Bautista Bru de Ramón in the royal museum in Madrid, from a set of fossil bones found in Alluvial deposits near Buenos Aires in Spanish South America. In 1796 a copy
of this still unpublished engraving was sent to the Institut in Paris, where the young Georges
Cuvier identified the animal as a giant sloth, named it *Megatherium*, and concluded that it was
probably extinct (modern research has given it a bipedal posture, up on its hind legs).

living species. Soon after his arrival in Paris, the Institut de France (which had likewise replaced the old royal Académie des Sciences) was sent engravings of a huge skeleton recently assembled in Madrid from fossil bones found in Spanish America. Cuvier compared the bones with those of living mammals from around the world. He claimed sensationally that the animal was closest to the living sloths and anteaters, far smaller mammals that he later classed as "edentates." This implied that what he named the *megatherium* ("huge beast") was probably extinct, for reports of such an enormous animal would surely have reached Europeans living or working in South America, if it was still alive. Like the "Ohio animal," it was a striking example of the way that natural history had become global in its sources, although still almost exclusively European in the scientific interpretation of its materials.

At almost the same time, Cuvier strengthened his argument by analyzing in detail the fossil bones and teeth of the mammoth, comparing them with those of living elephants. Luckily for him, the Muséum acquired relevant new specimens, just at the right moment, as cultural loot from the Netherlands (recently conquered in France's Revolutionary wars). Cuvier confirmed that the Indian and African elephants were distinct species. Much more importantly, he also claimed that the mammoth was distinct from either. He argued that the differences were as great and as consistent as those between, say, goats and sheep; they could not be attributed merely to the effects of age, sex, or environment. This knocked the bottom out of Buffon's claim that finding the remains of tropical mammals in Siberia was evidence of gradual global cooling. It also undermined the alternative idea that the carcasses had been swept there from the tropics, by a Flood in the form of a mega-tsunami. If the mammoth was in fact a quite different species, it might have lived where its bones are found, well adapted to the very cold climate that still prevails there. This was confirmed shortly afterwards by the discovery of mammoth skeletons buried in frozen ground in Siberia, preserved with thick woolly fur. The facts of comparative anatomy, Cuvier concluded, "seem to me to prove the existence of a world previous to ours, destroyed by some kind of catastrophe." He had been reading Deluc before arriving in Paris, and the echo of the older savant's concept of a "former world,"

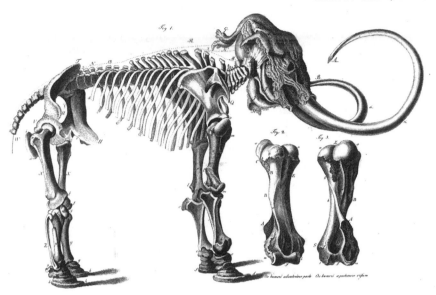

**FIG. 5.2** The complete skeleton of a mammoth found in 1799 in frozen ground in Siberia and brought to St. Petersburg, where this reconstruction was published in 1815. On the skull are scanty remains of the thick woolly fur that had originally covered the whole animal. This implied—as Cuvier had already suggested—that the species had been well adapted to the arctic climate of northern Siberia, rather than the tropical climates in which Indian and African elephants now live: mammoths could no longer be used as evidence for a gradually cooling Earth or for a violent Flood or mega-tsunami. (Two enlarged views of the huge thigh bone or femur, about a meter in length, are also included, making good use of the expensive copper plate from which the engraving was printed.)

separated from the present world by a drastic natural "revolution," was unmistakeable.

Cuvier then studied the "Ohio animal" and concluded that it was in fact so distinct from either elephants or mammoths that it deserved to be put in a new genus, which he named "*mastodon*" (alluding to the breast-shaped protuberances on its huge teeth). He went on to claim that the bones of fossil rhinoceros found with the mammoths in Siberia, those of large fossil bears found in caves in Bavaria, and many other fossil mammals from other Alluvial deposits around the world, were all distinct from their living counterparts (what he was studying was, in modern terms, the Pleistocene megafauna). The Institut, recognizing the exceptional importance of this research, published an appeal from Cuvier to naturalists and fossil collectors to send him further specimens, or at least accurate drawings of them. A generous international response—even at the height of what was in effect the world's first global war—enabled him vastly to enlarge his database of relevant specimens, and he published a steady stream of scientific papers analyzing one fossil mammal after another, comparing and contrasting each with its nearest living counterparts.

With a shrewd sense of scientific strategy, Cuvier focused his research on the remains of *large* land animals, because these were the species whose descendants—if any survived as "living fossils"—would be hardest to miss. Even if they lived in remote parts of poorly explored continents, large animals would be unlikely to have escaped notice; even if they had not yet been seen alive by naturalists, they were likely to feature in reports by hunters and trappers, or in tales told by indigenous peoples. Cuvier recognized that his case for the reality of extinction could be made only in terms of probability. The greater the number of fossil species that he could show were distinct from any similar living species, the more *probable* it became that they were truly extinct (Cuvier's work was published fully in 1812, the same year as an equally great work on mathematical probability by his older colleague Pierre-Simon Laplace, which gave the notion of probability both publicity and prestige). His case would be cumulative; and in the event it was, as he published more and more of his detailed analyses of fossil bones of all kinds.

One of Cuvier's colleagues at the Muséum remarked approvingly that he had sprung up "like a mushroom" on the scientific stage. Despite his youth, he was soon one of the most prominent savants in Paris, which in turn, despite the long years of war, was unquestionably the center of the scientific world. His unrivaled knowledge of comparative anatomy enabled him to reconstruct the skeletons of fossil mammals, even when what was available was just a lot of separate fossil bones (contrary to later legend, he did not claim to be able to reconstruct a whole animal from a single bone, but only to identify—in favorable cases—what *kind* of animal it had belonged to). His deep understanding of the relation between function and structure in animal bodies then enabled him to infer, with almost equal confidence, how these

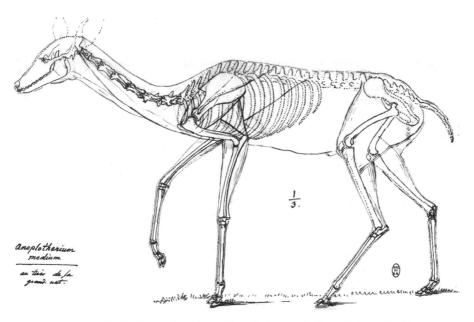

*Anoplotherium medium*

*sa taïs de la grand. nat.*

**FIG. 5.3** Cuvier's "resurrection" of one of the extinct mammals (*Anoplotherium medium*) from the Gypsum formation of Montmartre (then just outside Paris). Using his unrivaled knowledge and understanding of the comparative anatomy of a wide range of living mammals, Cuvier not only reassembled the whole skeleton from many scattered fossil bones, but also reconstructed the outline of the body and its likely posture, and even added its eyes and ears. Yet he never published this and other similar drawings, perhaps because he feared his scientific colleagues would regard them as unacceptably speculative.

fossil mammals had lived and moved and had their being. Alluding to the biblical prophet Ezekiel's vision of the valley of dry bones, he claimed to be bringing his fossil bones back to life, at least in the mind's eye; and, as he pointed out wryly, without the benefit of a divine trumpet. He was "resurrecting" animals as lively as those in the zoo next to his house in the grounds of the Muséum. In effect, Cuvier enriched Deluc's concept of a "former world" by populating it with a rapidly expanding zoo of previously unknown mammals, many of spectacular size and all of them likely to be extinct. Even if further "living fossils" were to turn up (as they did, though rarely), it became increasingly implausible to use them as a general explanation of the contrast between fossil and living species. Extinction now had to be taken into account as a real feature of the natural world. Fossil species could therefore be treated as reliable witnesses to a past that had been distinctly different from the present. The deep past really was a foreign country.

## THE EARTH'S LAST REVOLUTION

In due course Cuvier reprinted his papers in four volumes of *Researches on Fossil Bones* (*Recherches sur les Ossemens Fossiles*, 1812). Right at the start of its lengthy introductory essay, which was based on earlier lectures for the educated public in Paris, he described himself as "a new kind of antiquary": new, because his work was focused on nature's relatively unexplored "monuments" of fossil bones. These he could "restore," reconstructing the animals in order to trace "the ancient history of the Earth." The metaphors he borrowed from human history were almost as pervasive as Soulavie's a generation earlier. Cuvier was a new and more powerful advocate for turning traditional descriptive "natural history" into a true *history* of nature (in the modern sense of the word).

Cuvier drew a vivid analogy with the more prestigious sciences of astronomy and cosmology; he dedicated his work to Laplace, his patron in Paris and the greatest cosmologist of the age. He aspired to "burst the limits of time" by making the pre-human history of the Earth reliably *knowable* to human beings tied to the present, just as cosmologists had already "burst the limits of space" by making the laws of nature governing the motions of the Solar System known to human beings tied to one small planet. It was not the timescale itself that needed to be burst: Cuvier, like other savants, assumed it was almost unimaginably greater than that of human history (astronomers at this time could not detect any stellar parallax, which proved that interstellar space must likewise be almost unimaginably vast). What certainly was at issue was Cuvier's claim that the consistent differences between fossil mammals and their living counterparts were due to a relatively recent episode of mass extinction, and that the fossils were therefore relics of a sharply distinct "former world" beyond the reach of human historical records.

However, it could be objected that these species might have survived into the human period and then, like the dodo, have become extinct by human agency. To counter this, Cuvier adopted the role of a Classical scholar and reviewed a vast range of ancient Greek and Ro-

We admire the power by which the human mind has measured the movements of the globes [i.e., the planets], which nature seemed to have concealed forever from our view; genius and science have burst the limits of space, and observations interpreted by reason have unveiled the mechanism of the world. Would there not also be some glory for man to know how to burst the limits of time and, by observations, to recover the history of this world and the succession of events that preceded the birth of the human species?

FIG. 5.4 The analogy between cosmology and the new science of the Earth—Deluc's "geology"—as set out (originally in French) in the *Preliminary Discourse* to Cuvier's great *Researches on Fossil Bones* (1812). The "limits" that had been or could be "burst" were not those of the mere magnitudes of space and time, but rather the human capacities to *know* what in each case was beyond direct experience; Cuvier was alluding to Laplace's great work on "celestial mechanics," which was regarded as having perfected Newton's cosmology.

man descriptions and images of the animals known at that time. Apart from obviously mythical creatures, he identified them all as being in fact species still known alive. The only other possible explanation for the contrast between living and fossil species—if the "living fossil" argument was discounted—had recently begun to be championed by one of Cuvier's senior colleagues. Jean-Baptiste de Lamarck, who was the Muséum's expert on molluscs and other lowly animals (later named "*invertebrates*"), argued that animal species were ultimately unreal or arbitrary units, because all organic forms were, by their very nature, continually in flux. Given enough time, any one species would slowly and naturally *transmute* into another (using this contemporary word, rather than the modern one "evolve," will help to distinguish Lamarck's ideas from those put forward later by Darwin and others). Lamarck argued that if fossils differed from living species it was because they had transmuted in the meantime; if they were identical, it was because there had not yet been enough time for any transmutation to be detectable. In either case, Lamarck denied that any species had ever gone extinct unless by human agency.

Cuvier countered Lamarck's argument by taking the sacred ibis of the ancient Egyptians as a test case. Its mummified remains had been collected by the savants attached to Napoleon's military expedition to Egypt during the Revolutionary wars. Cuvier compared the ancient ibis with museum specimens of living birds, and identified it as a species still flourishing in the Nile valley. Its anatomy had not changed perceptibly in some three thousand years. He conceded that this was an extremely short time in relation to the Earth's vast history; but if all species were in fact continuously transmuting, even a small lapse of time should show some slight change (astronomers used similar reasoning when they plotted the lengthy orbits of the outer planets by extrapolating from accurate observations over a much shorter period). And anyway Cuvier considered Lamarck's slow transmutation impossible even in principle. In his view, and that of most other zoologists, each species was an "animal machine" in which every organ was integrated with all the others to make possible a particular mode of life (in modern terms, most animals are very precisely adapted to specific environments). Any slow transmutation, changing the anatomy away from this viable state, would therefore make the species unfit to survive, long before it could attain a new viable state as a new species well adapted to a new and different way of life. So Cuvier believed that species were real natural units, necessarily stable in form and habits unless and until they were wiped out by some sudden disruption of their environment.

If then the fossil species had been completely replaced by the present lot, yet not by the transmutation of one into the other, what had caused the extinction of the first set and where had the second set come from? Cuvier assumed that a major natural "revolution" or "catastrophe" must have killed off the older lot. The ancient continents might have collapsed suddenly beneath the sea, as Deluc had argued; or, as others had suggested, there might have been a mega-tsunami caused by a mega-earthquake, just like those that on a smaller scale had notoriously destroyed the city of Lisbon in 1755. To explain where the new species had come from, Cuvier proposed an ingenious thought-experiment, which involved possible *future* migrations of Australia's recently discovered marsupial mammals and Asia's well-known placental ones, as their respective continents were flooded or emerged. Somewhat conveniently, this shelved or postponed the question of accounting for the *origins* of the new species that seemed to have replaced the extinct ones. But *causal* questions about either extinctions or origins were tangential to Cuvier's research: he was trying to reconstruct nature's *history*. On both points his goal was to establish the historical reality of the events; working out their physical causes was quite another matter.

Cuvier was not smuggling into his argument any hint that either extinctions or origins of species were anything other than purely *natural* events. His theorizing was not driven by a covert desire to bolster any kind of modern-style creationism. He had been brought up as a Lutheran and his cultural loyalties remained with the small Protestant (mainly Reformed or Calvinist) minority in France; much later he acted as its official liaison with the state, and helped to protect its civil rights. But his personal religion seems to have been formal or even perfunctory; his pious daughter, before her early death, was praying for his true conversion. The atheists among the Parisian savants regarded him as their ally, until he dismayed them by claiming that the Earth's recent "revolution"—with its mass extinction of his fossil mammals—was none other than the Flood. Primarily, however, this was just a part of his attempt, following Deluc, to use the biblical event to tie early human history on to the tail end of the Earth's own history, and thereby to integrate the two.

Cuvier's endorsement of Deluc's ideas on the historical reality of Noah's Flood was embedded in a wider argument that went far beyond the biblical story. Again he adopted the role of a historical scholar: he searched reports of all known ancient literate cultures, as far afield as China, noting that many of them included similar traditions of some kind of catastrophic inundation around the dawn of their histories. In reviewing them all, he put the Genesis story in first place because he thought it the oldest of its kind (and of course it was also the best known to his readers). He cited a leading German orientalist and biblical critic for the likely date of the Genesis text, and he treated it simply as a surrogate for the still older (and as yet undeciphered) records of the ancient Egyptians, which Moses could have learnt about during the Jews' long exile in Egypt. All this was hardly the reasoning of a biblical literalist. Anyway, Cuvier concluded that all these multicultural records—however obscure and even garbled they might be—were compatible with the opinion of Deluc and other earlier savants, that the Earth's recent "revolution" could not have been more than a few millennia in the past.

This made it the decisive boundary-event not only between the present and the former world, but also—almost—between the human world and the pre-human. Cuvier conceded that there must have been some "antediluvial" human beings in existence before the catastrophe struck, or there would have been no memory of the event among its survivors and therefore no subsequent records. But he assumed that they would have been confined to a few limited regions (perhaps just Mesopotamia and one or two other spots) and therefore of no great importance on a global scale. If so, his worldwide zoo of vanished species was unlikely to have been wiped out by the activities of human beings (as the dodo had been, on one small island). To confirm this, Cuvier used his unrivaled expertise in comparative anatomy to review the many claims that human bones had been dug up along with those of his extinct animals. He systematically eliminated all of them. Either they were not human at all, as in the case of Scheuchzer's much earlier "man a witness of the Flood" (which Cuvier identified as a giant salamander!); or they were not found unambiguously in the same deposits as the other bones and were probably not of the same date; or they were recent in origin and not really fossils at all. He concluded that no genuine human fossils were known: the "former world" destroyed in the Earth's recent "revolution" had been an almost completely *pre-human* world. This confirmed what many savants already suspected: humans had appeared on stage at a very late moment in the immensely long drama of the Earth's total history.

Cuvier's "resurrection" of a spectacular zoo of extinct mammals, and his interpretation of it as a vanished "former world" that had been destroyed by some kind of geologically recent "revolution" or "catastrophe," had a huge impact not only on savants internationally but also on the public that heard him lecture in Paris or read his more accessible published work (in French or in its many translations). His persuasive prose, and powerful position at the center of the scientific world, ensured that in the early 19th century his vision of the Earth as a product of nature's own *history*, while far from being original to him, came to be taken for granted by savants and widely absorbed by the educated public around the Western world.

## THE PRESENT AS A KEY TO THE PAST

However, Cuvier and his supporters did not have it all their own way. The idea of an exceptional event in the Earth's recent past, let alone

a unique one, was unacceptable to some savants. Desmarest, for example, had found no evidence for such an event in his detailed reconstruction of the physical history of Auvergne. He had concluded
that the ordinary process of erosion by rain and rivers had alone been
responsible for an unbroken and gradual modification of the topography. In the new century, in vigorous old age, Desmarest continued
to insist on the total adequacy of such processes. Hutton's Edinburgh
friend John Playfair made a similar case, but with more ambitious
generality, in his *Illustrations of the Huttonian Theory of the Earth*
(1802; the illustrations were textual, not pictorial), which after Hutton's death made that theory more palatable to a new generation; a
French translation later made Playfair's work more widely accessible.
Playfair used Hutton's prose style—which is not really as obscure as he
claimed—to justify his own rewriting of the theory. He played down
Hutton's pervasive deistic arguments, based on nature's intelligent design, and almost airbrushed them out of sight. Since he was primarily
a physicist and an astronomer, it is not surprising that he emphasized
instead the unchanging laws of nature. These obviously underlay what
Deluc had called "present causes," the ordinary physical processes observably at work in the present world. So it is also unsurprising that
such processes figured prominently in Playfair's argument. He claimed
that "present causes" were sufficient to explain all the traces of what
had happened in the deep past. There was no need to bring in any
other causal agents or to suggest any unusual or catastrophic events.

Savants had in fact taken it for granted, at least since the time of
Steno and Hooke, that this kind of reasoning was the most effective strategy for understanding the past history of the Earth: it was
later expressed in the aphorism, familiar to modern geologists, that
"*the present is the key to the past.*" Since Deluc's own term for observable present processes was *causes actuelles*, this method of research
was later termed "*actualism*" ("actual" in this sense of "current" or
"present-day" is now almost obsolete in English, so "actualism" has yet
to become familiar to modern Anglophone geologists, though it has
long been used by those in the rest of the world). However, major dis-
agreements arose over the adequacy of "actual" or present processes
to account causally for *all* events in the Earth's unobservable deep
past. Cuvier, for example, was quite explicit about this. He claimed
that no known present process was adequate to account for his mass

extinction, and that an exceptional "revolution," unparalleled in the observable present world, must have been responsible for it. But this did not make the event any less natural than everyday processes such as erosion, or any less grounded in the physical laws of nature. It just implied that certain natural processes might have operated occasionally with such exceptional intensity that they were almost different in kind. For example, as already mentioned, the "revolution" that had caused the mass extinction might have been a mega-tsunami, a hugely scaled-up version of the kind of devastating natural event that had been witnessed and recorded in Lisbon half a century earlier.

However, in addition to advocating the actualistic *method*, Playfair also promoted Hutton's *theory* that the Earth works as a cyclic or steady-state system, from and to eternity. This was far more controversial. It entailed rejecting the growing evidence not only for an event of exceptional intensity in the relatively recent past, but also for the overall directional character of the Earth's even deeper history (a point to which the next chapter will return). To most other savants, even if they were sceptical about Cuvier's identification of the "revolution" with the Flood in Genesis, a blanket rejection of the possibility of *any* kind of unusual event in the deep past seemed highly questionable or even perverse. At least in the short run, Playfair's perspective found few supporters.

## THE TESTIMONY OF ERRATIC BLOCKS

Evidence for the reality of a recent and violent "revolution" came not only from Cuvier's fossil bones but also—and still more strikingly—from huge blocks of rock scattered on the surface of the ground in several parts of Europe. Such blocks were often composed of rock that was quite different from the bedrock on which they lay, and in many cases it could be identified with what formed the bedrock in some area tens or even hundreds of miles away. These *"erratic blocks"* were too large and too numerous to be attributed to human activities; yet they were also so large, and had erred and strayed so far, that it seemed that only a physical "revolution" of almost unimaginable magnitude could have had the power to shift them. They had to be seen to be believed. When an anonymous Scottish reviewer of Cuvier's work (probably Playfair) was publicly sceptical about the reality of any such event, a

**FIG. 5.5** A huge erratic block of Primary granite, lying on bedrock of Secondary limestone on the slopes of the Jura hills above the city of Neuchâtel in Switzerland, some 100 km (60 miles) from where the same variety of granite was found as bedrock, high up in the Alps near Mont Blanc. This particular erratic became well known among naturalists after Leopold von Buch used it in 1811 as his prime example in a major paper read at the Academy of Sciences in Berlin. This sketch was drawn by the English geologist Henry De la Beche, when in 1820 he visited this famous spot on his way to do fieldwork in Italy, where he saw many more large erratics. The puzzle was nothing if not international.

Genevan savant commented that such doubts could easily be dispelled if the Scotsman would only come and see for himself the huge erratics near his city: they had evidently been transported—somehow—all the way from the high Alps near Mont Blanc.

In fact the Genevan erratic blocks belonged to just one of several long trails of erratics on the northern flanks of the Alps, which were explored and described by the leading Prussian savant Leopold von Buch (who was in the region primarily to survey the mineral resources of the then Prussian territory around Neuchâtel). He admitted that he was utterly perplexed by them, but the least implausible explanation seemed to be in terms of a sudden and violent mass of water such as a mega-tsunami. A few years later von Buch's puzzlement was alleviated by reports of a tragic disaster in one of the Alpine valleys. A huge barrier of rock and ice had ponded back a large lake near the top of the Val de Bagnes (on the same spot there is now a concrete dam for a reservoir for hydro power). One day in 1816 the natural dam suddenly burst and released a massive and violent flow of turbid mud, which overwhelmed several villages and transported large blocks of rock at high speed for many miles down the valley. It seemed a striking small-scale model for a far larger event that might have transported huge erratics, as von Buch had traced them, all the way from the high Alps across the Swiss plain and up on to the flanks of the Jura hills to the north. This authentic "present cause"—in modern terms, a sort of sub-aerial "turbidity current"—showed that the alleged "revolution" could have an intelligible natural explanation. The date of such an event was impossible to estimate, though it was obviously very recent in relation to the rest of the Earth's long history; but whether it could be equated with the biblical Flood was a separate issue.

However, it was more difficult to explain the trails of distinctive erratics that von Buch and others later traced for hundreds of miles over relatively low ground in Sweden and Finland, right across the Baltic Sea and then over the plains of Russia and northern Germany. One huge erratic block, for example, had been used in St. Petersburg as a pedestal for the famous equestrian statue of its founder Peter the Great; another near Berlin had been fashioned into a huge granite bowl to decorate an open space in the city center. Savants could hardly ignore such accessible evidence for an exceptional event of some kind. It was suggested that the blocks might have been given buoyancy by being embedded in winter ice floes before they were swept across northern Europe, but this lessened only slightly the immensity of what seemed to have happened.

The Scottish savant James Hall was another of those who argued for the reality of a recent violent event, and also explored its possible cause, while leaving aside the question of its relation to the biblical Flood. As a young man, Hall had been a friend of Hutton's in Edinburgh; but unlike Hutton he had seen for himself the Alpine erratics on the Jura hills, while returning from a cultural Grand Tour in Italy, and he recognized what they implied. Back home, he was equally impressed by the surfaces of bedrock scratched with parallel grooves, and other linear features, that he found around Edinburgh. He thought that if masses of small stones, carried in suspension in turbid water, had swept violently over the area, they might have scratched the underlying bedrock as they passed by. To account for the much

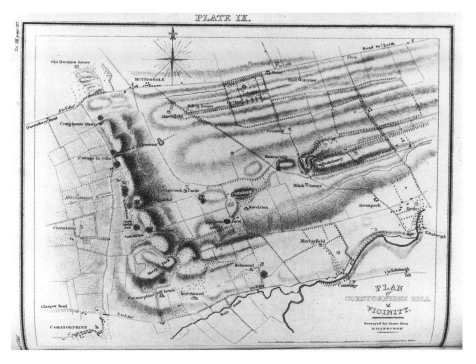

**FIG. 5.6** James Hall's map of Corstorphine Hill (now in the suburbs of Edinburgh), showing the linear topography that he interpreted as the trace of a "*diluvian wave*" or transient mega-tsunami that had flowed violently from west to east. He suggested that stones swept over the hill had scratched the hard bedrock in the same direction: the dark spots mark localities that he listed as notable "specimens" of such scratched bedrock, preserved in this outdoor museum of nature's antiquities. His paper, read at the Royal Society of Edinburgh in 1812, was in effect an invitation to other savants to go and judge the evidence for themselves.

larger erratic blocks, he too thought that they might have been given buoyancy by being embedded in ice. And to explain the event itself, Hall suggested that any sudden upheaval of the ocean floor, if on a large enough scale, would generate a mega-tsunami and produce just the right effects (as usual the Lisbon tsunami provided a smaller-scale model). Although he admired Hutton's grand theory, Hall saw no reason why the upheavals of the new continents that were required to maintain the Earth's steady state might not have been sudden events. So he argued that the specific event he was reconstructing was just the most recent example of an effect that had punctuated the indefinitely long history of the Earth. There was nothing special about it except its relatively recent date; occasional violent "revolutions" were, he claimed, an intrinsic part of the way the Earth works.

## BIBLICAL FLOOD AND GEOLOGICAL DELUGE

Combined with Cuvier's spectacular extinct animals, erratic blocks provided the most powerful evidence for the reality of a drastic "revolution" at some point in the relatively recent past. Cuvier's claim that it had happened around the dawn of human history was particularly attractive and persuasive. If it could be identified as the Flood described in Genesis and other ancient records, it would tie human history into the geologists' novel account of the Earth's deep and pre-human history, forming an impressive single narrative.

However, most savants recognized that the case for the reality of an exceptional episode in the Earth's recent history was distinct from the case for its having happened recently enough to be equated with the story of Noah. The physical episode itself (assuming it was a real event) was often called the *"geological Deluge."* Whether it had also been the *biblical Flood* recorded in Genesis, or a far earlier event that was pre-human in date, was a separate question. In what became known as the *"diluvial theory"*—quite distinct from the much earlier ideas of Woodward, Scheuchzer, and others—claims for the physical reality of the geological Deluge did not necessarily imply support for the historical reality of Noah's Flood, and in some cases were explicitly opposed to that idea. In what follows, *"Deluge"* will be used in this sense, to denote the supposed physical event, whether or not it was believed to have also been the *"Flood"* recorded in Genesis (savants at the time were less consistent in their use of these words, and anyway those using English were at an advantage in having both at their disposal).

However, in contrast to von Buch, Hall, and others, many naturalists did explicitly endorse Cuvier's claim that a massive Deluge—now traced so sensationally in Switzerland and northern Europe—had been so recent that it could indeed be identified as the biblical Flood. This was the slant that Robert Jameson, Edinburgh's professor of natural history, gave to Cuvier's famous introductory essay when he had it translated into English. He presented Cuvier's work as a new "Theory of the Earth" that would help to refute the atheistic implications of

Hutton's eternalism, although in fact Cuvier himself scornfully repudiated *all* such Big Pictures. Jameson recruited the immense prestige of Cuvier's science to bolster the cultural authority of political conservatism in Britain in its alliance with traditional religion (his editorial slant has, ever since, given Cuvier an undeservedly negative reputation in the English-speaking world). Jameson claimed that despite sceptical voices to the contrary—not least Playfair's, in their home city of Edinburgh—the historical reliability of the Genesis narrative had now been vindicated by Cuvier. And so in Britain, in contrast to the rest of Europe, the relation between geology and the interpretation of Genesis became entangled with contentious political and cultural issues (a point to which this narrative will return).

Among the readers of Jameson's edition of Cuvier's work was the young Oxford don William Buckland, who had just been appointed to give a course on mineralogy. He chose to include in his lectures the exciting new science of geology. He rapidly brought himself up to speed, not least by studying Cuvier's full *Fossil Bones* (like other educated Britons he could read French fluently, just as non-Anglophone scientists now have to read English as a matter of course). He accepted Cuvier's dating of the geological Deluge, but he followed Deluc in focusing solely on its compatibility with the biblical story rather than the diverse multicultural records that Cuvier had reviewed. He highlighted the evidence for the reality of the Deluge because he saw in it an opportunity to convince his Oxford colleagues that geology, far from being a threat, was a science that deserved to be taught at their university, the intellectual center of the Church of England. If the geological Deluge could be identified as the biblical Flood, geology would confirm the historical reliability of the story in Genesis and thereby reinforce the authority of the Bible in general.

Almost on his own doorstep Buckland soon found new evidence for the Deluge, to add to what von Buch, Hall, and many others had already described. With the assistance of his future wife Mary Morland—who became a notable exception to the usual exclusion of women from geological fieldwork—Buckland mapped the Alluvial gravel deposits around Oxford, and noticed that they contained some distinctive pebbles that were certainly not local in origin. He found that these gravels (with these pebbles) extended not only down the valley of the River Thames towards London but also upstream, up over the Cotswold Hills and down on to the Midland plain to the north, where he traced the pebbles to their source. Reconstructing the *history* that had produced this puzzling distribution, he inferred that the little pebbles, like the huge Alpine erratic blocks, could act as nature's antiquities: they marked the track of a massive "*diluvial current*" that had apparently swept across this part of England in the geologically recent past. Geologists soon began to refer to such deposits collectively as "*diluvium*": this distinguished them from a more narrowly defined "*alluvium*," a term now reserved for the still more recent or post-diluvial deposits. The diluvium included not only gravels and large erratic blocks, but also widespread deposits of "*till*" or "*boulder clay*," full of angular chunks of rock of all sizes, which was puzzling because nothing like it was known to be forming in the present world.

Unlike von Buch and Hall, Buckland combined his reconstruction of a physical "diluvial current" with Cuvier's focus on the fossil evidence for the same event. On his first Continental tour—after Napoleon's final defeat at Waterloo in 1815 at last made such travel practicable for Englishmen—Buckland made a point of seeing for himself the famous caves in Bavaria, from which masses of fossil bones had already been retrieved. Cuvier had identified them as the bones of bears, though larger than any of the living species and probably extinct. But it was not clear whether they had been living in the caves before the Deluge struck or had been swept in from elsewhere, perhaps very far away, during that event.

Luckily for Buckland, a cave discovered in 1821 nearer home enabled him to resolve this puzzle and use fossil bones to take the reconstruction of the deep history of the Earth to a new level. Paradoxically, the bones found in Kirkdale Cave in Yorkshire, in northern England, provided decisive evidence precisely because the cave was much smaller and far less impressive than those in Bavaria. Using Cuvier's methods, the bones were identified as those of a wide variety of animals, ranging in size from water rats to mammoths. But the cave was far too small for carcasses of the larger ones to have been swept in by the waters of the Deluge itself. Instead, Buckland found abundant evidence that the cave had been used as a den by a large extinct species of hyena, which must have scavenged the carcasses outside and then dragged bits of them into the small cave to enjoy at leisure. Using the standard actu-

VERTICAL SECTION OF THE CAVERN AT GAILENREUTH IN FRANCONIA.

FIG. 5.7 Buckland's illustration (published in 1823) of the already famous Gailenreuth cave near Muggendorf in Bavaria, showing abundant bones of fossil bears being excavated from deposits underlying the layer of stalagmite on the floor of the cave. The stalagmite, which was still being precipitated slowly from water dripping through the cave, acted as a "natural chronometer." Buckland argued that it proved that the bears had lived in the cave—or their carcasses had been swept in—long ago, before the start of the "present world."

alistic method, Buckland watched the eating habits of living hyenas in a zoo and found identical tooth marks on the fossil bones; likewise he identified other objects found among the bones as the distinctive excrement of bone-chewing hyenas. So the cave and its bones could be used as a window onto the "former world" immediately *before* the Deluge struck. As in the Bavarian caves, the evidence for this vanished world had been sealed in by the overlying layer of stalagmite that had accumulated subsequently. The very modest thickness of this deposit acted like one of Deluc's "natural chronometers." It proved that the geological Deluge that wiped out the world of the cave hyenas and

FIG. 5.8 Buckland crawling into Kirkdale Cave and finding the extinct cave hyenas alive and well, feasting on the bones of animals large and small, and as startled to see him as he is to see them. Entering the cave with "the light of science" in hand, he is also time-traveling back into the antediluvial world that his careful research has made reliably knowable. This caricature—jokey in form yet serious in meaning—accompanied a similarly jokey poem by his friend and former Oxford colleague William Conybeare, celebrating Buckland's achievement in having opened this "spy hole" into the Earth's deep history. Copies of this lithographed broadsheet (1823) circulated widely among geologists in Britain and beyond; for example, one was sent to Cuvier in Paris.

mammoths had been, in geological terms, very recent. It was recent enough, Buckland claimed later, for it to be identified as none other than the biblical Flood.

Buckland reconstructed a whole fauna of herbivores, carnivores, and scavengers that had been living in the area around Kirkdale Cave. His "hyaena story," as he called it, brought back to life not just individual animals, as Cuvier had, but a set of interacting species (in modern terms, an ecosystem). His contemporaries saw it as a spectacular vindication of Cuvier's aspiration to "burst the limits of time" by making the pre-human past reliably *knowable* to humans confined to the present. Buckland's research showed, even more clearly than Cuvier's, that it was indeed possible to construct a conceptual time-machine. His kind of thorough investigation could transport humans into the deep past: not just as imaginative science fiction but as a *history* of nature as firmly grounded in solid evidence as the scholarly human history that it emulated. This was the conclusion of Buckland's lengthy report on Kirkdale Cave to the Royal Society, which gave him its highest award for it. His subsequent book on *Diluvial Relics* (*Reliquiae Diluvianae*, 1823; only the title was in Latin) then made his work more widely accessible. Having traveled extensively through France and Germany to see for himself as many bone caves as possible, in addition to excavating others in his own country, he described a mass of evidence for the Europe-wide geological Deluge that he now identified explicitly as the biblical Flood. Echoing an Oxford colleague's *Sacred Relics* (*Reliquiae Sacrae*), which had reviewed the historical evidence for the foundational events of Christianity, Buckland's title underlined the equally historical character of his reconstruction of this enigmatic earlier event.

In view of Buckland's intellectual and personal investment in equating the geological Deluge with the biblical Flood, he might have been expected to welcome any new evidence that human beings had in fact been contemporary with the extinct mammals: any human bones found with the "diluvial" fauna would be genuinely those of "a man a witness of the Flood." But Buckland remained as deeply sceptical as Cuvier about the authenticity of any alleged human fossils; he shared Cuvier's scientific caution about all such claims. He himself found a human skeleton close to a mammoth skull, beneath the floor of a cave on the south coast of Wales. But since it was preserved quite differently he claimed that it was a burial dug in Roman times into far older diluvial deposits; the skeleton became famous as the "Red Lady." However, she did not weaken Buckland's confidence in the equation between Deluge and Flood, because he believed that at the time of the event human beings had probably still been confined to regions far from Europe (possibly just Mesopotamia). He probably thought that a Britain inhabited by lots of large wild beasts had been no place for humans to flourish; it was only after the former had been wiped out

FIG. 5.9 Buckland's picture of Paviland Cave on the coast of south Wales, published in 1823: this part of his section through the cave showed a human skeleton (right) close to the bones of extinct mammals. Although so near a mammoth skull, the bones were differently preserved, stained red, and became known as the "Red Lady." Buckland believed that she was buried, probably in Roman times, in a grave dug into much older deposits dating from before the geological Deluge. He therefore denied that this was evidence that human beings had been living in Britain alongside mammoths before the Deluge struck. (Modern research has confirmed Buckland's two-phase interpretation, although the skeleton, now identified as a young man, is dated as Palaeolithic, and the other fossils as still earlier Pleistocene).

that the latter could safely spread into Europe. The Deluge therefore remained an event that marked a sharp boundary between the human world and an almost completely non-human "antediluvian" world.

In the early 19th century, Buckland's careful research provided some of the most persuasive evidence for the reality of an exceptional Deluge event in the geologically recent past, which could have been responsible for the mass extinction that Cuvier's equally careful research had first disclosed. And all this was powerfully reinforced by the fieldwork of von Buch, Hall, and many others, who traced the paths of erratic blocks and other relics of "diluvial currents" such as scratched bedrock and boulder clay or till. In fact the diluvial theory went from strength to strength, as such features were traced ever more widely: on both flanks of the Alps, across vast tracts of northern Europe and even across the northern United States and Canada. Whether this "geological Deluge" had been recent enough to be equated with the Flood story in Genesis was much more controversial, though Cuvier's multicultural review of early human records provided suggestive evidence that it might have been. It also showed that those who argued for the identity of Deluge and Flood were not necessarily motivated— as Buckland certainly was, at least initially—by the hope that the new scientific evidence would reinforce the authority of the biblical record. The diluvial theory was first and foremost a *scientific* idea, which made good sense of most of the evidence available. The date of this geological Deluge in relation to human history, and specifically to the biblical Flood, was a separate issue, although obviously an important one.

This chapter has described how the naturalists who now routinely called themselves *"geologists"* were agreed that the Deluge event— which might or might not have been the biblical Flood—had been, in geological terms, very recent. The evidence for it was confined to the Diluvial deposits which, like the still more recent Alluvial ones, overlay even the uppermost and youngest of all the Secondary formations. The next chapter describes how, still in the early 19th century, geologists tried to unravel what had happened during the Earth's deeper history, long *before* the geological Deluge.

# 6

## Worlds Before Adam

### BEFORE THE EARTH'S LAST REVOLUTION

Most of Cuvier's fossil bones had been found in Alluvial deposits such as the gravels that he attributed to the Earth's geologically recent "revolution" (which might be faintly recorded in the biblical Flood story among many others). But some bones came from underlying Secondary strata, and specifically from the gypsum deposits being quarried just outside Paris (to make plaster for building work). These bones were more of a challenge than those of mammoths and mastodons, because Cuvier judged that they came from mammals much less like any known living species. If he was to "burst the limits of time" by making the pre-human past fully knowable, he needed to understand the place of these animals in the Earth's history. That entailed knowing the place of the gypsum in the total pile of formations, so his fossil anatomy needed to be combined with the quite different science of geognosy. As a few earlier naturalists had realized, geognosy could provide a framework for reconstructing "nature's chronology," but only if its static description of three-dimensional rock *structures* could be interpreted as evidence of the Earth's dynamic *history*.

Cuvier therefore formed a working partnership with Alexandre Brongniart, a mineralogist who was the new director of the state porcelain factory at Sèvres just outside Paris. In the first years of the 19th century they undertook extensive fieldwork in the Paris region (Brongniart was also looking for new sources of ceramic materials). They adopted the well-tried methods of geognosts; Werner himself visited Paris around this time, which may have helped. They used the surface outcrops of the rocks, and underground quarries and boreholes, to work out the three-dimensional structure of the pile of formations around Paris. They built on what the great chemist Antoine-Laurent Lavoisier had discovered, before his career was tragically cut short by the guillotine during the Revolution. They confirmed that Paris lay at the center of a shallow bowl or "basin" formed by the distinctive Chalk formation, which was found in boreholes deep beneath Paris itself, but rose to the surface in the surrounding countryside.

This "Paris Basin" proved that the Chalk, far from being the uppermost of the Secondary formations, was overlain by a thick pile of still younger formations, which included sandstones, clays, and limestones as well as the distinctive gypsum. In 1808, at a meeting of the Institut de France, Cuvier and Brongniart summarized their survey with a "geognostic map" showing the distribution of all these formations as they would be seen if the concealing cover of soil and vegetation were stripped away. This kind of map was not new: one of Werner's colleagues at Freiberg had published a similar map of Saxony, and Lavoisier and his colleagues had made detailed maps of many parts of France, plotting the distribution of distinctive rocks and minerals. However, Cuvier and Brongniart, like Soulavie and a few other earlier naturalists, found that they could recognize specific formations in the field, not only by the different kinds of rock and their associated minerals, but also by their distinctive fossils. Geognosy could therefore be *enriched* by adding common fossils such as mollusc shells to the other criteria that were used routinely to distinguish different formations and to trace their outcrops across the land.

However, Cuvier and Brongniart enriched their geognosy still further, to help reconstruct the Earth's *history*. For example, the distinctive Coarse Limestone—the attractive *calcaire grossier* of which much of Paris was built (and still is, in the older parts of the city)—contained fossil shells that were clearly similar to those of living marine species.

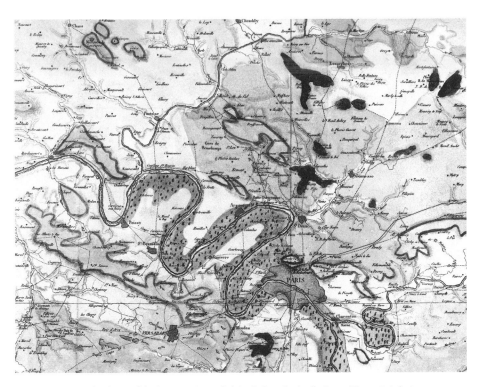

FIG. 6.1 A part of the "geognostic map" of the Paris region by Cuvier and Brongniart, first displayed in 1808 at the Institut de France and published in 1811. It shows (in color in the original) where the different rock formations are found at the surface: the Gypsum formation, in which some of Cuvier's most puzzling fossil bones were found, is represented by the dark patches and loops (on the tops and flanks of some of the hills); the stippled area along the meandering valley of the River Seine represents the far younger river gravels, with the bones of mammoths and other extinct mammals. Maps of this kind became a standard tool in the hands of geologists everywhere, helping them to visualize the geognosy or three-dimensional structure of the rocks in a particular area.

They were natural antiquities that were evidently relics of a vanished sea, which must have extended across much of the Paris region. This was unsurprising: most common fossils in most Secondary formations looked more or less similar to animals now living in the sea. What certainly was surprising was that in some of the other Paris formations all the fossil shells were closely similar to those of shellfish known to live only in freshwater. The pile of formations taken as a whole therefore implied that there had been a repeated but irregular *alternation* of marine and freshwater conditions. The Paris region seemed to have been alternately a sea and a lake (or freshwater lagoon), before the recent "revolution" finally carved valleys in all these rocks and deposited Alluvial gravels in them.

This was an impressive reconstruction of an eventful history of the region, during the obviously very long time since the Chalk at the base of the Parisian pile was first deposited. Cuvier and Brongniart were interpreting the formations not just as a pile of rocks but as a historical sequence of alternating periods of marine and freshwater conditions in the deep past. So this was a geognosy *doubly* enriched—and thereby transformed—by the use of fossils to reconstruct the past *history* of the region. And Cuvier and Brongniart took their historical reconstruction still further. They found in the field that some of the junctions between the marine and freshwater formations (with their respective fossils) were quite sharp. They interpreted this as a sign that the changes from marine to freshwater conditions, or from freshwater back to marine, had been correspondingly sudden. In other words, these junctions were evidence of *repeated* "revolutions": natural changes less violent and dramatic than the puzzling diluvial event in the more recent past, but nonetheless further proof that the Earth—at least in the Paris region—had had a complex and eventful history. And it seemed to have been a history as irregular, contingent, and unpredictable (even in retrospect) as the recent human history of France itself, with its bewildering sequence of coups d'état during the years of Revolution and the further eventful years of war under Napoleon.

This detailed analysis of the Paris Basin by Cuvier and Brongniart became the most influential example of how the static three-dimensional structures described by geognosts could be transformed into a dynamic *history* of a specific part of the Earth. It also drew attention to these relatively recent Secondary formations lying above the Chalk. Their sheer scale and variety had not been generally appreciated; and the detailed description of their fossils highlighted their contrast with the older Secondary formations, from the Chalk downwards in the pile. Their animals and plants were much more like living organisms than the fossils found in the older formations: for example, there was a conspicuous absence of ammonites and belemnites, which were often abundant in the older rocks. Cuvier urged geologists to give high priority to a closer study of these younger formations and

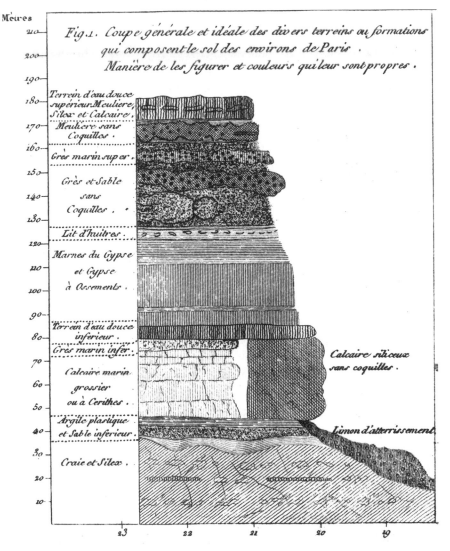

Fig.1. Coupe générale et idéale des divers terreins ou formations qui composent le sol des environs de Paris.

Manière de les figurer et couleurs qui leur sont propres.

210—
200—
190—
180— Terrein d'eau douce supérieur Meulière, Silex et Calcaire.
170— Meulière sans Coquilles.
160— Grès marin super.
150—
140— Grès et Sable sans
130— Coquilles.
120— Lit d'huîtres.
110— Marnes du Gypse
100— et Gypse à Ossements.
90—
80— Terrein d'eau douce inférieur.
70— Grès marin infer.
60— Calcaire marin
50— grossier ou à Cérithes.
40— Argile plastique et Sable inférieur.
30—
20— Craie et Silex.
10—

Calcaire siliceux sans coquilles.

Limon d'atterrissement.

23    22    21    20    19

**FIG. 6.2** The "general and ideal section" through the formations of the Paris Basin, published by Cuvier and Brongniart in 1811. They interpreted the pile of rocks as the historical record of a repeated alternation between marine conditions (marked *marin*, and a bed of *huîtres* or oysters) and freshwater (*d'eau douce*) conditions. The Gypsum formation with fossil bones (*Gypse à ossements*) is near the middle of the pile, with the Coarse Limestone (*Calcaire grossier*) below it and the Chalk (*Craie*) at the base. The formations are depicted as if they were all exposed in an imaginary cliff on the side of a valley eroded through them, with Alluvial gravel deposited much later at the foot of the cliff. (The two formations occupying the same position in the pile were puzzling at the time; in modern terms it was a case of two *facies*, deposited in different environments during the same period.) The section depicted a typical thickness of about 150 m (500 feet) for all the formations above the Chalk.

their fossils: being relatively recent they might provide a key to the still earlier history of the Earth.

In fact, several geologists elsewhere in Europe were already making a start on describing these younger formations, but Cuvier's huge prestige gave their work a heightened importance. Giovanni-Battista Brocchi, who was in charge of the new natural history museum in Milan (modeled on the great Muséum in Paris where Cuvier was working), studied the "Subapennine" formation in the foothills of the Apennine Mountains. Its abundant fossil shells, like those that Lamarck was describing from the Coarse Limestone around Paris, were similar to those of molluscs living in present seas. Brocchi called all such formations *"Tertiary,"* to distinguish them from the still older formations that continued to be called Secondary. The name was soon adopted by geologists everywhere (and continues to be used by their modern successors). Brocchi's work confirmed that the Tertiary formations and their fossils represented a distinctive major period in the Earth's total history.

Around the same time the London surgeon James Parkinson (whose medical work later gave his name to the debilitating disease) described the Tertiary formations in a "London Basin," which like the Parisian one was underlain and bounded by the Chalk formation. And Thomas Webster, an artist and architect who worked in London at the newly founded Geological Society, described a similar "Isle of Wight Basin" on the south coast of England. After Brongniart sent samples of his Parisian freshwater fossils to London (evading Napoleon's wartime blockade), Webster recognized that there were identical fossils in his strata, and indeed a similar alternation of marine and freshwater formations, suggesting a similar history.

The Isle of Wight Basin (later renamed Hampshire Basin) also revealed something else that was new and important about this relatively recent period. The island's cliffs showed that the Chalk, and even some of the overlying Tertiary formations, had been tilted spectacularly into an almost vertical position. Similar huge *"folds"* had long been known in the Alps and elsewhere, but they affected much older Secondary or Primary rocks, so they could safely be attributed to extremely remote times in the Earth's early history. In contrast, the Isle of Wight showed that here at least the Earth's crust had undergone a dramatic major upheaval relatively recently. This implied a much more dynamic kind of

SCRATCHELLS BAY AND THE NEEDLES, I.W.

FIG. 6.3 Thomas Webster's view of the Needles, at the west end of the Isle of Wight off the south coast of England. The white limestone of the thick Chalk formation (its strata highlighted by bands of black flints) has been tilted into an almost vertical position on the north (left), though less so to the south (right). In other engravings, Webster showed that in Alum Bay (further to the left) the overlying Tertiary formations were also vertical. This proved that the Earth's crust had here been folded on a massive scale, as recently as the Tertiary period in the Earth's history, and subsequently eroded. This dramatic scene was published in 1816 in Henry Englefield's lavishly illustrated book on the "picturesque beauties, antiquities, and geological phenomena" of the island; the title reflected the wide range of readers that the book was designed to attract, including not only Webster's fellow members of the Geological Society but also antiquarians and local residents.

history for the Earth than most geologists—apart from some of Hutton's followers such as Hall—had imagined hitherto. It suggested that they might be able to "burst the limits of time" and make this history knowable through the evidence not only of fossils but also from that of the structure of the Earth itself (in modern terms, its tectonics).

However, Brocchi's Italian research showed that there was yet more to be learned from the less dramatic evidence of fossil shells, in the light of detailed knowledge of *living* molluscs, which was readily available thanks to the contemporary fashion for collecting pretty shells from around the world. Brocchi found that about half the species in the Subapennine formation were unknown alive, and he inferred that most of them must be extinct (a few might survive somewhere as "living fossils," as yet undiscovered, but this had become implausible as a *general* explanation of apparent extinctions). But the other half of the fossil species were known to be still alive, many of them not far away in the Mediterranean. This raised an issue of outstanding importance. Cuvier's idea of a *mass* extinction, to explain his fossil mammals, could not apply to these Tertiary molluscs, because some species had disappeared while others had not.

Brocchi's volumes describing and illustrating his Subapennine shells (*Conchiologia Fossile Subapennina*, 1814) included a major essay on the question of extinction, in which he suggested that in the case of these marine animals extinction might instead be a *piecemeal* process. He followed Cuvier (and most other naturalists other than Lamarck) in treating species as real natural units, well adapted to particular ways of living and therefore unchanging in form for as long as they existed at all. But he suggested that each species might have a limited life-span, becoming extinct for intrinsic reasons rather like the death of an individual in old age. This in turn implied that species might appear on the scene in a similarly piecemeal fashion, by some process analogous to the birth of an individual. If so, the *history* of molluscan life as a whole would be rather like the history of a human population: it would change slowly but continuously in composition, as species (like individual human beings) appeared and disappeared from time to time. If furthermore the average life-span of species varied between one group of organisms and another (just as the life-span of individual animals varies enormously between, say, insects and mammals), and if the rate of turnover was for some reason faster in mammals than in molluscs, this could account for the otherwise puzzling fact that all the fossil mammals found in Tertiary formations appeared to be extinct whereas many of the fossil molluscs were not.

Brocchi's conjectures offered a perspective on the history of life in sharp contrast to Cuvier's. They suggested a picture of piecemeal and quite gradual change, rather than one of occasional abrupt change. Nature's revolutions might be analogous to the smooth and regular "revolutions" of the planets around the Sun (as in the title of Copernicus's celebrated book *De Revolutionibus*) rather than the violent political "revolutions" that had convulsed both Europe and North America in the recent past. Brocchi's and Cuvier's models seemed to apply rather well in their own spheres, for marine and terrestrial animals respectively. But further research on the fossils of the Tertiary formations soon began to broaden the matter into more fundamental questions about the character of the Earth's history. As Cuvier had anticipated, the Tertiary period, by linking the present world to the even deeper past, could be the key to understanding the whole.

## AN AGE OF STRANGE REPTILES

When Cuvier embarked on his ambitious research project on fossil bones, he aimed to identify and describe those of *all* "quadrupeds": that is, not only mammals but also reptiles (including those later classed as amphibians) and birds. He soon established that many of the bones found in formations underlying the Tertiary ones—those from the Chalk downwards in the pile, now redefined as "Secondary" in a narrower sense than before—were the bones of reptiles, not mammals. One example, already famous, was the "Maastricht animal" found in the Chalk formation near that Dutch city. Cuvier, with his unrivaled knowledge of comparative anatomy, confirmed that it was not a toothed whale, nor even a crocodile, but a huge marine lizard; it was later named the *mosasaur* ("lizard of the Maas or Meuse," the river on which the city lies). Equally spectacular, although far smaller in size, was what Cuvier identified as a *flying* reptile and named the *ptéro-dactyle* ("wing-fingered"); it was a reptilian equivalent (now termed a pterosaur) of the birds and the bats. Other isolated bones from other Secondary formations looked like those of crocodiles, and Cuvier was sure that they too were not those of any mammal. So he suggested provisionally that the Secondary formations had been deposited at a time when quadrupeds were represented *only* by reptiles, and that mammals (and birds) had not appeared until later, in the Tertiary period. This was a first hint of a possible *history* of quadruped life; even perhaps a history that could be called "progressive," since human beings, universally regarded as the "highest" of all mammals, seemed to have appeared even more recently, with no authentic fossils at all.

   Cuvier's suggestion was strengthened sensationally by further fossils found in England during and soon after the years of war. Some of the bones that had been ascribed to crocodiles turned out—when almost complete skeletons were found—to belong to much more pe-

culiar creatures. The best specimens came from the Secondary formation that had long been known as the *Lias*, composed of alternating
layers of limestone and shale. They were found not by geologists but
by *"fossilists"* who made a precarious living from finding fossils that
they could sell to collectors, tourists, and other people of higher social
class. One such fossilist, Mary Anning of Lyme Regis on the south
coast of England, was unusual in being a woman but also in having
an exceptionally sharp "eye" or flair for finding fine fossils (modern
heroic myth-making about her has obscured the fact that this envi-
able gift was, and is, rarely combined with the skills and knowledge
needed to interpret the fossils scientifically once they are found). Anning's first great find was of a huge fish-shaped reptile that had neither
a crocodile's legs nor a fish's fins, but flippers like those of a dolphin.
It was named the *ichthyosaur* ("fish-lizard"), which aptly expressed its
puzzling character. A decade later, Anning found a rather similar but
very long-necked reptile that was given the equally apt name *plesiosaur* ("almost-lizard"). Its anatomy was even more peculiar and unparalleled: so much so that the cautious Cuvier warned that it might
have been assembled fraudulently by some deceitful fossilist. But it
was analyzed rigorously by William Conybeare, Buckland's former
colleague at Oxford, who, using Cuvier's own anatomical methods,
demonstrated at the Geological Society that it was certainly genuine.

Conybeare also emulated Cuvier by "resurrecting" the plesiosaur
in the mind's eye; he suggested how its strange anatomy could have
enabled it to live effectively as a marine fish-eating carnivore. Meanwhile, Buckland was analyzing in a similar fashion some of the other
fossils that Anning was finding in the same Secondary formation.
For example, inspired by his own achievement with the hyenas' den
at Kirkdale, he analyzed some distinctive objects found with the fossil bones and claimed they were the reptiles' excrement. They contained the distinctive scales of a fossil fish found in the same rocks,
so Buckland inferred who had eaten whom, and reconstructed the
food chain; his apparent fascination with fossil feces enhanced his
carefully cultivated reputation as a learned eccentric. All this research
was memorably summarized in a print made by the English geologist
Henry De la Beche (he had a Norman-French family name). Its Latin
title, *Duria antiquior* ("a more ancient Dorset"), suggested a parallel
with scholarly reconstructions of the Classical world, but this was a

FIG. 6.4 A reconstruction of life at the time that a specific Secondary formation (the Lias) was deposited, based on fossils found at Lyme Regis in the county of Dorset on the south coast of England. This scene from deep time, depicting "a more ancient Dorset," was drawn in 1830 by Henry De la Beche. Most prominent is an ichthyosaur chopping the long neck of a plesiosaur, which is extruding its feces; another ichthyosaur has just caught a belemnite (reconstructed as a cuttlefish); another, of a longer-jawed species, is swallowing a fish. A larger fish is capturing a lobster; pterodactyls, one of them just caught on the wing by a plesiosaur, are flying overhead; a crocodile and a turtle are on the shore; and in the background are palms and cycads, indicating a tropical climate. Some of the finest specimens of these fossils had been found by the local fossilist Mary Anning, for whose benefit the print was sold; but the analysis on which the reconstruction of the ecosystem was based was not her work but largely that of Buckland.

view of a specific period in the Earth's deep pre-human history. It was the first full-fledged example of a new kind of image, a "scene from deep time," but it was modeled on the well-established artistic genre of scenes from *human* history (including those from biblical history, such as Scheuchzer's). This new scientific genre, depicting reconstructions of *nature's* history, proved highly effective in conveying to geologists—and later to the general public—what could be inferred about the character of the Earth and its life in the deep pre-human past (and of course it remains essential in modern museum displays and, with computerized animation, in movies and TV programs). Nothing could illustrate more clearly how Cuvier's aspiration to "burst the limits of time" was being fulfilled.

## THE NEW "STRATIGRAPHY"

Any such reconstruction of life and its environment, based on fossils found in a specific formation, would obviously benefit from a precise knowledge of the place of that formation in the total pile and hence its place in the Earth's total history. Just as Cuvier had recognized that his giant extinct mammals from the Alluvial deposits were much more recent than the more peculiar mammals from the Tertiary formation of the Paris gypsum, so it was clear, for example, that the mosasaur from the Chalk was much more recent than the even more peculiar reptiles from the Lias: the Chalk formation was known to lie far above the Lias. A more precise history of what seemed to have been an age of reptiles would therefore depend on a more detailed knowledge of the pile of Secondary formations.

This kind of geognostic knowledge was becoming increasingly available at just this time. The practical value of ordinary fossils such as mollusc shells had been noticed not only by Cuvier and Brongniart but also by William Smith, an English mineral surveyor. Smith had begun to compile a geognostic map of the rock formations in England and Wales a few years before the two Parisians began theirs (Brongniart may have seen a copy of Smith's draft map, and been inspired by it, when he visited London in 1802 during a brief interlude of peace). But Smith had great difficulty getting his much larger and more complex map published, and it was not generally available until 1815, four years after the one of the Paris region. This led to bitter arguments, often chauvinistic in tone, about which map deserved credit for being the first of its kind. In fact, however, the practical value of maps showing the distribution of rocks and minerals had been recognized by several geognosts back in the 18th century; as mentioned earlier, even the French map had had precursors.

However, Smith's huge map was certainly a huge achievement, since he had surveyed the whole of England and Wales single-handed. It covered a much wider area than the Parisian map; it traced the outcrops of a larger number of distinct formations (which Smith idiosyncratically called "*Strata*"); and he made much greater use of what he called "*characteristic fossils*" to recognize them as he traced them across the land. But Smith's map did not change the world (as ignorant modern heroic myth-making has claimed) or even the world of geology. Its size and beauty ensured that it became a much admired model for later maps of the same kind. Yet it remained implicitly, as the Parisian one was explicitly, a *geognostic* map, although the robustly insular Smith would not have accepted that un-English term. It displayed the *order*—his own favorite word—of the English formations in their three-dimensional structure. Smith used his "characteristic fossils," along with rock types and other well-tried criteria, to help identify and define the successive formations, but he did not use them to reconstruct a history of the Earth or even of England. He himself tacitly acknowledged this when he coined a new word to define what he was doing: "*stratigraphy*" expressed his intention simply to *describe* the "strata" or formations. Stratigraphy was an appropriate word (and one that remains essential in modern geology) for what has here, up to this point, been called "enriched geognosy" (in the German lands where geognosy had first been developed, stratigraphy continued for many decades to be known as *Geognosie*).

In the early 19th century stratigraphy became the staple scientific work of most geologists. Their most common kind of publication was a detailed description of the formations (and their fossils, if any) in some specific area. This was usually illustrated with a geological map and often also by sections showing the pile in profile: the combination enabled geologists to visualize in the mind's eye the three-dimensional structure of the Earth's crust in that area. Although the sequences varied in detail from one region to another, they could often be "*correlated*" by recognizing the same fossils in equivalent formations, even when the types of rock were not closely similar. Smith's insistence on the supreme value of "characteristic fossils," as the criterion to be preferred above any other for correlation, turned out to be generally justified. But all this was still stratigraphy, or geognosy enriched by the use of fossils; it was not in itself a reconstruction of the Earth's history.

One of the most influential summaries of stratigraphy was the *Outlines of the Geology of England and Wales* (1822), which was largely compiled by Conybeare. He was well aware of the possibilities of reconstructing the Earth's history, having around the same time "resurrected" the strange plesiosaur and designed the famous caricature of

Buckland in a den of extinct hyenas. But his book had a quite different, stratigraphical goal, and drew extensively on Smith's work. He summarized the formations in England and Wales, and their likely equivalents beyond Britain, in order from the top down: a useful order for unraveling geognostic structure, although the very opposite of the order that represented the Earth's history. Most of his book was devoted to the many and varied Secondary formations from the Chalk downwards, past the Lias and ending with what he defined as "*Carboniferous*" ("coal-bearing") formations. The book stopped at this point, because the formations even lower in the pile were as yet poorly explored. The sequence of Secondary formations in England and Wales turned out to be more complete and more consistently full of fossils than in most other parts of Europe of comparable size. Thanks largely to Conybeare's work, geologists everywhere came to treat Britain's geology as a valuable standard of reference, at least for the Secondary formations.

Over the next two decades the different parts of the pile of Secondary formations were given names that were eventually accepted internationally by informal agreement among geologists (and they remain familiar to their modern successors worldwide). At the top of the pile, the Chalk, and some formations below it with rather similar fossils, became known as "*Cretaceous*," from the Latin word for chalk itself. Below them were formations that were named "*Jurassic*," after the Jura hills on the Franco-Swiss border where they were well exposed (the Lias was near the base of these formations). Below them in turn was a set of formations that was called "*Triassic*" because across much of central Europe it had a tripartite character: two sandstone formations were separated by a distinctive limestone (in England the limestone was missing and the other rocks were known as "New Red Sandstone"). Still lower in the pile were some rather similar sandstones and another limestone, with underground deposits of salt, also widespread around Europe; the name for them all that was eventually adopted was "*Permian*," from the city of Perm at the foot of the Ural mountains far away in Russia, where they were even better displayed. They in turn were above the "*Carboniferous*" formations already mentioned. These included not only the Coal formation itself—of supreme economic importance in Europe's burgeoning Industrial Revolution—but also some thick underlying formations; the lowest of these was a distinctive "Old Red Sandstone" that was taken as the base of the Secondary formations as a whole.

These Secondary formations often rested directly on Primary rocks such as granite and gneiss, with no fossils at all. But in some regions they were underlain instead by rocks, such as slates, for which Werner had earlier proposed the name "*Transition*": they were transitional between Secondary and Primary, in both position and character, having just a few fossils. The successful charting of the pile of Secondary formations made the underlying Transition ones an obvious challenge. In the 1830s, they in turn began to be unraveled, when it was found that in some regions they were as straightforward in structure and as full of fossils as the Secondary formations above them. One such region was the Welsh Marches (the border country between England and Wales). Here the London geologist Roderick Murchison—who had married an heiress and had an ample budget for his research—defined a set of "*Silurian*" formations, named after the ancient British tribe that had lived there in Roman times: the antiquarian analogy between human history and the Earth's own history remained both potent and useful. Underlying them in turn, but before any Primary rocks were reached, were formations that Adam Sedgwick, Buckland's counterpart at Cambridge, named "*Cambrian*" after the Roman name for Wales itself; they had very few fossils, and most of those were poorly preserved and not clearly distinct from the Silurian ones. (This led to a bitter dispute between Murchison and Sedgwick, which was resolved only much later when the "*Ordovician*" group familiar to modern geologists was inserted irenically between a redefined Silurian and Cambrian).

Not in this straightforward order was one other major group of formations, which was named "*Devonian*" after the English county of Devonshire. This was born out of what was called at the time the "*great Devonian controversy.*" It was resolved only when geologists across Europe agreed that the distinctive Old Red Sandstone was—as a very puzzling anomaly—of the same age as formations in other regions (Devonshire among them) of a totally different character with totally different fossils. The reason for this anomaly remained for some time quite obscure, but anyway the Devonian could confidently be inserted between the Silurian and a more narrowly defined Carboniferous.

What is obvious about all these names for major parts of the pile of Secondary and Transition formations is that they referred either to distinctive kinds of rock or to regions where the relevant rocks were well exposed: the names were all based on stratigraphical (or geognostic) criteria. Each came to be known as a "*system*," by which was meant a distinctive group of formations with an equally distinctive set of fossils. Once the Devonian controversy was settled, their structural or geognostic *order* in the pile of formations was unambiguous and uncontroversial.

## PLOTTING THE EARTH'S LONG-TERM HISTORY

However, the same names soon came to be used to define the "*periods*" of time during which the relevant formations had been deposited: those comprising the "Jurassic system" had been deposited during a "Jurassic period." This illustrates how the practice of stratigraphy, although not itself historical, could provide a framework for reconstructing the Earth's history. It could provide not just a glimpse of the deep past through an occasional and fortunate "spy-hole"—the metaphor that Conybeare had applied to the scene of the hyenas' den at Kirkdale—but a continuous narrative account of the long-term history of the Earth and its life. When in 1836 Buckland summarized for the general educated public (and also his colleagues) what geologists had so far discovered, he was able to anchor his account in some two decades of extraordinarily fruitful international research in stratigraphy.

Buckland was also able to use all this new stratigraphy to portray what was becoming an increasingly clear picture of the history of life on Earth, as revealed by the sequence of fossils. This "*fossil record*" (as it was later called) started with the Transition formations and went through all the Secondaries and then the Tertiaries, ending with those in the Diluvial and Alluvial deposits. But it was more useful, as Cuvier (and Steno and Desmarest before him) had recognized, to analyze this in the first instance in reverse order from present to past, from the familiar to the unfamiliar. Starting then from the present, it was clear that the most recent past—immediately before the apparent geological Deluge—had been a time of spectacularly large mammals (in modern terms the Pleistocene megafauna), but that most of them were quite closely similar to living species. The Tertiary era before that time had been characterized by many organisms such as molluscs that were, in

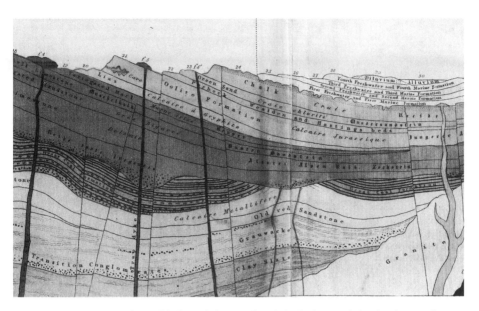

**FIG. 6.5** A small part of the huge ideal section through the Earth's crust (colored in the original) drawn by Webster and published by Buckland in his *Geology and Mineralogy* (1836). At the very top of the pile is the modern "Alluvium," and below it the "Diluvium" attributed by Buckland (and most other geologists) to the recent "geological Deluge." Below them are the Tertiary formations, shown with alternating marine and freshwater formations as in the Paris Basin. Next come the Secondary formations, from the Chalk down to the strongly-marked "Great Coal Formation" and underlying "Old Red Sandstone." Below the latter are Transition rocks such as "Grauwacke" and "Clay Slate" and finally the Primary rocks, marked here as "Granite." There are also "igneous" rocks, originally forced upwards as hot fluids from a great depth, some reaching the surface as volcanic lavas. The trilingual names (English in Roman font, French in italics, German in Gothic) express the "correlations" made across Europe, and the international character of stratigraphical research. The relative thicknesses of the formations were thought likely to be roughly proportional to the amounts of time that the rocks represented: the "modern world" of human life, represented only by the Alluvium, was assumed by Buckland and his contemporaries to cover an extremely brief recent period in comparison with the Earth's deeper, complex, and unimaginably lengthy history. This section was designed a few years before "system" names such as Jurassic and Carboniferous came into general use among geologists.

the same way, similar to their modern counterparts; some, as Brocchi had shown, were even identical. But the Tertiary mammals, such as those that Cuvier had first analyzed, were much less like living forms; further fossil finds heightened the sense that even the Tertiary era had been quite strange in character.

Moving further back in time, it was increasingly clear that most of the Secondary formations had been deposited during an age of reptiles. The marine mosasaur, the ichthyosaur and plesiosaur, and the fly-

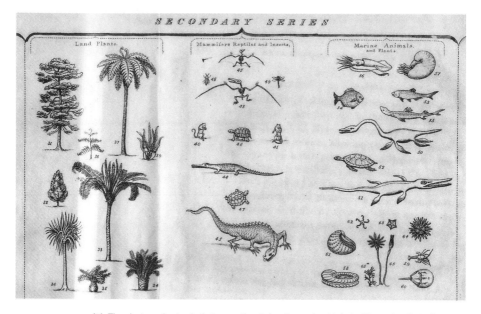

FIG. 6.6 The plants and animals that were alive during the era in which the "Secondary Series" of formations were deposited: one of the sets of tiny drawings attached to Buckland's great stratigraphical section (1836). Where possible the fossils were reconstructed to suggest their appearance in life; many had been portrayed in De la Beche's famous scene of "a more ancient Dorset." Among the set of land animals (middle column) are the huge herbivorous *Iguanodon* reptile and two tiny mouse-like marsupials that were the earliest known mammals. The organisms were not, of course, drawn at a uniform scale.

ing pterodactyl were soon joined by other reptiles. Buckland claimed that the *megalosaur* ("giant lizard"), which he based on fossils found near Oxford, was a terrestrial carnivore; and the English country surgeon Gideon Mantell described an *iguanodon* ("teeth like the iguana," a far smaller living lizard), which at Cuvier's suggestion he interpreted as another terrestrial reptile, but a huge herbivore (both these were known only from a few teeth and bits of bone, so reconstructions of them were far more conjectural than those of the other fossil reptiles). In 1841 the zoologist Richard Owen—who earned the nickname of "the English Cuvier"—assigned them on anatomical grounds to a new group of extinct reptiles that he named the *dinosauria* ("terrible lizards"); further discoveries suggested that dinosaurs had been confined to the Jurassic and Cretaceous periods.

The inference that this had been an age of strange and varied reptiles—marine, terrestrial, and aerial; carnivorous and herbivorous—

was hardly dented when some very rare fossils of very small mammals were found in the same Jurassic formation as the megalosaur. In fact they were just what might have been expected, for Cuvier identified them as tiny opossum-like marsupials. So this more "primitive" kind of mammal seemed to have existed well before the ordinary and more "advanced" placental kinds made their appearance in the Tertiary period. This reinforced the growing sense that the quadrupeds had had a total history that was "progressive" in character. But vertebrate animals were not the only strange creatures from the Secondary era. Among the fossil molluscs were highly varied ammonites and abundant belemnites, both assumed to be totally extinct (naturalists had in effect given up hope of finding any as "living fossils"); even the more ordinary molluscs were almost all distinct from any living species. And there were many other exotic marine animals; for example, sea-lilies (stalked crinoids) were quite abundant, in contrast to their extremely rare occurrence as "living fossils."

Still further back in the history of life, the successful unraveling of the older Secondary and Transition formations was beginning to disclose an era with animals yet more strange in character. In the Carboniferous and Devonian formations, the oldest of the Secondaries, no fossil reptiles were found, indeed no quadrupeds at all, but only fish. Fossil fish of all ages were being described in detail by the young Swiss naturalist Louis Agassiz, whose *Researches on Fossil Fish* (*Recherches sur les Poissons Fossiles*, 1833–43) complemented Cuvier's great work on fossil quadrupeds. Agassiz claimed that in the Carboniferous and Devonian formations the fish—all of them complex in structure and some very large in size—were of kinds that were either totally extinct or at least had become rare in later periods. And since none were found in the still older Silurian formations, there was a growing suspicion that the seas of that period might have been without any vertebrate animals at all.

Silurian invertebrates, however, were abundant and varied, though strangely unlike later forms. Most striking of all were the "*trilobites*." Like the earliest fish, these were complex animals. They were somewhat similar to living animals with jointed external skeletons (arthropods), such as crabs and lobsters, yet they were quite distinct from any of them. Trilobites were abundant and varied in Silurian and Devonian formations, and there were a few in the Carboniferous, but

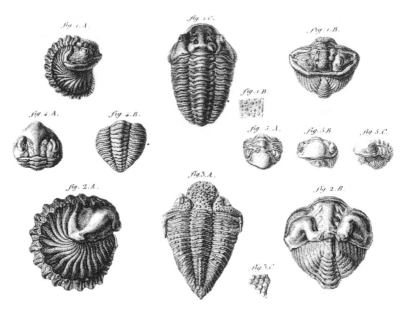

**FIG. 6.7** Trilobites of the genus *Calymene*, as pictured in Brongniart's volume on trilobites (1822). In the 1830s the Transition formations in which these fossils were found were defined by Murchison as being part of his *"Silurian"* system, which made them some of the oldest fossils then known. They were obviously complex animals with a jointed external skeleton and large compound eyes, able to curl up tightly—presumably for protection against predators—rather like living woodlice and armadillos. They were far from being the crude and simple organisms that would have been expected as the earliest forms of life, on Lamarck's theory of "transmutation" or evolution.

none beyond the Permian. So trilobites seemed to characterize the Transition era and early part of the Secondary just as ammonites and belemnites characterized the later part of the Secondary era. Finally, before even the Silurian period, all that could be found in the Cambrian formations were a very few trilobites and the shells of a very few other invertebrate animals: it seemed that they brought the fossil record close to the beginnings of life itself.

This complex record of animal life was paralleled by that of plant life. Just as, among vertebrate animals, the major groups seemed to have appeared in succession—first fish, then reptiles, then mammals and birds—so the major groups of plants appeared in the fossil record in a somewhat analogous sequence. At the earliest end of the fossil record, no trace of land plants of any kind was found in the Silurian formations: it looked as if plants at this time must have been confined to marine algae or seaweeds. The succession of later plant life was plotted by Brongniart's son Adolphe, a rising scientific star, in his *[Natural] History of Fossil Plants* (*Histoire des Végétaux Fossiles*, 1828–37); this was an ambitious (and never completed) work that, like Agassiz's on fish, emulated Cuvier's on quadrupeds. Again like Agassiz, the younger Brongniart used the new stratigraphy to chart the total *history* of plant life. The first and oldest set of plant fossils were those found abundantly in the Coal formation (their decayed remains seemed to have formed the coal itself). They were flowerless plants (Cryptogams) in great variety, many of them tall trees that were in striking contrast to the modest size of most of their living counterparts the ferns, horse-

FIG. 6.8 A scene in a forest of the Carboniferous period, as reconstructed for August Goldfuss's *Fossils of Germany* (*Petrifacta Germaniae*, 1826–44; only the title was in Latin, the text in German). The plants were all identified as flowerless Cryptogams, related to living ferns, horsetails and clubmosses, but growing into tall trees with massive trunks. Fossil leaves were rarely found attached to trunks, and it was uncertain which belonged with which. So the scene was shown—ingeniously—cut off above a few meters, with the fronds broken off and strewn on the ground. Fish are swimming in the creek (bottom right), while some of the invertebrate animals found in marine strata of about the same age are shown thrown up on the shore (bottom center).

tails, and clubmosses. In the younger Secondary formations, cycads and conifers (gymnosperms) made their appearance and were indeed abundant. Only in the Tertiary period had flowering plants (angiosperms) also become abundant and varied. (Set against this orderly history of plant life, the bitter Devonian controversy first erupted when De la Beche reported finding large fossil land plants of Carboniferous species in strata of apparently Silurian or even Cambrian age; but further fieldwork eventually showed that the relevant strata were in reality Carboniferous, so the glaring anomaly was eliminated.)

## A SLOWLY COOLING EARTH

The fossil record of both animals and plants clearly indicated a history that was linear and *directional*. In both cases it could also be interpreted as "progressive," in that "higher" kinds of organisms seemed to have appeared in succession in the course of time: mammals later than reptiles or fish, flowering plants later than the flowerless kinds. How then was this strongly directional character of the history of life to be understood?

One possible clue was the apparently tropical look of many of the plants and animals from the earlier periods. For example, in what are now the cool temperate latitudes of northern Europe there were coral reefs in the Jurassic and also much further back in the Silurian. And there were shells almost identical to the pearly nautilus, which was known to live around the East Indies (now Indonesia); it had long been suggested that the rather similar and much more abundant ammonites might have been equally tropical in habitat. Still more striking was the evidence of fossil plants. Even the Tertiary formations around London contained the remains of plants similar to those now living in much warmer places than chilly England. Further back in time, De la Beche had good fossil evidence for depicting palm trees and cycads on the Jurassic shore of his "earlier Dorset." The fossil coal plants from still further back, in the Carboniferous period, were even more impressive in this respect, recalling the dense jungles and mangrove swamps of the present tropics. And coal deposits and fossil corals were reported from the high Arctic, by expeditions searching for a "northwest passage" through the maze of icy channels around the top of North America (modern ideas of continental "plates" slowly shifting in latitude, powered by plate tectonics, were of course unimaginable at this time).

The younger Brongniart linked all this evidence of formerly hotter global climates with what his Parisian colleagues the famous physicist Joseph Fourier and the geologist Louis Cordier had been suggesting, namely that the Earth must have cooled gradually from an incandescent origin. In essence, this was just what Buffon had proposed half a century earlier. But now it could be deployed on an implicitly vast timescale far beyond what Buffon had envisaged, and it could be linked to Laplace's influential "nebular hypothesis" that the planets had all condensed from a plume of intensely hot material emanating from the Sun. Most importantly, it could be supported by Fourier's latest and best mathematical physics of heat, and by Cordier's review of the latest and best measurements of the rise of temperature in mines (the geothermal gradient). In effect they argued that the Earth certainly had an "internal" or "central heat," which they thought could be interpreted most plausibly as a residual heat. To Brongniart and many other geologists, this geophysical theory (to use the modern term) made good sense of all the varied evidence that in the deep past the Earth's surface had been much hotter than it now is. If much of the heat had formerly come from the Earth's interior rather than from the Sun, it would explain why coal forests seemed to have flourished even at high latitudes: global climates might have been much more uniform, and less dependent on latitude, than they have since become. It could also explain why the fossil record petered out as it was traced back in time: the Cambrian period might have been the earliest in which the seas had cooled sufficiently to allow any life at all.

Adolphe Brongniart took this idea of a long-term environmental history of the Earth still further. He suggested that the sheer luxuriance of the coal forests of tree-ferns and giant horsetails and clubmosses might have been made possible by an early atmosphere in which the "carbonic acid" (carbon dioxide) necessary for photosynthesis had been much more abundant than it now is. Conversely, this might have delayed the appearance of the "higher" forms of animal life such as the mammals, which require plenty of oxygen. On this view even the atmosphere, no less than the solid Earth and the life at its surface, might have had its own deep history, and one that could in principle be reconstructed. Such large-scale theorizing encouraged

FIG. 6.9 "The Earth: Supposed to be seen from Space": the frontispiece of De la Beche's *Researches in Theoretical Geology* (1834). Like most of his visual images, it was drawn rigorously to scale, showing the very slight flattening of the globe—into an oblate spheroid—that was widely taken to be evidence of the Earth's originally fluid state. At this period a viewpoint from deep space was generally left to astronomers; De la Beche was one of the very few geologists who were using it to think in global terms about the Earth itself. His book expounded the directional theory that interpreted the history of the Earth in terms of a very slow cooling from a very hot origin, leaving it with residual heat in its deep interior.

at least a few geologists to begin to think again about the Earth in literally global terms, as one planet among others. This was a way of thinking that had generally been rejected, since early in the century: either as too speculative to have any place in the respectable new science of geology, or as falling outside its proper realm and better left to the astronomers.

Finally, the model of a gradually cooling Earth could, paradoxically, explain a feature of the Earth's history that seemed to be anything but gradual. The new stratigraphy made it possible to date—on the unquantified or *relative* timescale of the successive periods of the Earth's history—the times at which there had been major disturbances of the Earth's crust. These were often marked by local "*unconformities*" between older and newer sets of formations, like those that Hutton had used long before to claim there had been a "succession of former worlds." In each case the older set had clearly buckled, and its strata had then been worn down by erosion, before the newer set had been deposited on top. Extensive fieldwork all around Europe and even beyond it, by von Buch and many other geologists, showed that these episodes of upheaval had affected different regions at widely spaced intervals throughout the Earth's long history: the Isle of Wight, for example, had evidently been affected during one of the most recent episodes. The French geologist Léonce Élie de Beaumont, in a major article on "The revolutions of the surface of the globe" (1829–30), claimed that each of these occasional "*epochs of elevation*" had been marked by a huge buckling of the solid crust in some kind of mega-earthquake. He suggested that each had been caused when the rigid crust adjusted suddenly to a deeper core that was shrinking slowly and steadily as the Earth cooled down. In other words, a steady and literally underlying physical cause in the depths of the Earth could generate occasional "catastrophic" effects at the Earth's surface.

By the middle of the 19th century, this kind of reconstruction of the Earth's immensely long history was adopted by most geologists all across Europe, and also by the relatively few who were working beyond Europe in, for example, Russia and North America. They were agreed that the Earth seemed to have undergone broadly directional change, probably caused ultimately by its gradual cooling from an extremely hot origin to its present state. Animals and plants that were well adapted to its changing environments had appeared and disappeared accordingly, the "higher" forms being generally later in origin than the "lower." The whole fossil record was therefore not only linear and directional in character but also broadly "progressive," with the human species making its appearance only at the very end of the story. And this continuous change, although quite gradual for most of the time, seemed to have been punctuated occasionally by brief episodes of sudden and more violent change, which were assumed to have had equally natural causes, probably in the inaccessible depths of the Earth. However, this comfortable consensus among the world's geologists was disturbed, from at least three different directions. They are the topic of the next chapter.

# 7

## Disturbing a Consensus

### GEOLOGY AND GENESIS

The new science of geology embodied a novel and unfamiliar perspective on the Earth's own history. In particular, of course, it presented overwhelming evidence that human history had been preceded not just by one week of God's primal creative action but by an immensely lengthy and eventful history, all of it apparently pre-human, stretching back into an unimaginably remote past. It reduced the whole of known human history to a brief final scene in a far longer drama, and it made any naively "literal" interpretation of the Creation narratives in Genesis literally incredible.

However, this conclusion was far from stirring up the kind of acrimonious conflict between Science and Religion that modern atheistic fundamentalists (and some religious ones too) like to imagine. For a start, many of the leading scientific actors in this historical drama, particularly in Britain, were in public life ordained clergymen, and in private life sincerely devout Christians. Buckland and his Cambridge counterpart Sedgwick were typical in this respect. They combined the teaching of geology at the two English universities with religious duties at major Anglican cathedrals, and both were prominent national figures in each role. Another example was Buckland's former Oxford colleague Conybeare, who was not only one of the most intelligent geologists of his generation, but also a fine theologian and church historian who was later in charge of one of the cathedrals in Wales. Conybeare worked hard to introduce the insular English to the kind of biblical criticism long practiced in the rest of Europe; he and his theological allies believed that only a more scholarly approach to the Bible could save it from being dismissed, in a newly scientific age, as irrelevant or worse.

Among such savants, however, those who were geologists had to establish the validity of their new understanding of the Earth's history, and its relation to human history, against the background of a public culture in Britain that was not always receptive to their novel ideas. They had consciously forged a common sense of identity, *as "geologists,"* in London's Geological Society. When this was founded in 1807 it had been the world's first body of its kind (a similar French society followed in 1830 and others still later in other countries). Initially it had claimed to be committed above all to the primacy of observation, and particularly outdoor fieldwork, rather than the speculative "Theories of the Earth" that had characterized the previous century. In Britain, in the febrile political atmosphere of the long Revolutionary and Napoleonic wars, there had been a deep suspicion of any novel or radical ideas—scientific or otherwise—emanating from France. So the Geological Society had kept its collective head below the parapet, by emphasizing its apolitical goal of simply assembling plain and useful facts about the Earth. But its very success had soon made this policy untenable. Even the plainest of facts called for interpretation, which meant theorizing about their meaning and significance. This in turn had increasingly involved making historical reconstructions of the Earth's own past, which were bound to be compared with already existing ideas about the world's origins and early history. Paramount among these were of course those culled from the narratives of Creation and the Flood in the early chapters of Genesis.

In early 19th-century Britain, then, the members of the Geological Society worked in a context of a lively literate culture, in which their novel ideas were compared and often contrasted with traditional ways of interpreting the Bible. This wider public interest in geology had its origins early in the century and was enlivened by Jameson's promotion—or rather, distortion—of Cuvier's work. But it was further intensified in the 1820s in the wake of Buckland's efforts to demonstrate geological evidence for the biblical Flood, and his sensational and well-publicized "hyaena story" about the antediluvial world. In this context, the two-volume *Scriptural Geology* (1826–27) published by the Anglican clergyman George Bugg has often been taken in retrospect as symptomatic of a sharp conflict at this time between geology and Genesis: Bugg's work was subtitled uncompromisingly "Geological Phenomena Consistent Only with the Literal Interpretation of the Sacred Scriptures." But this reading of the history is much too simple and indeed deeply misleading.

Understandably, the leading geologists did tend to focus their attention and their criticism on works such as Bugg's, which originated outside their circle and explicitly challenged their own authority on matters of geology. They tended to exaggerate the threat such authors represented; they even cast themselves—however jokingly—as heroic figures in the mold of Galileo confronted by the Inquisition. The threat seemed greatest when the sharp boundary between themselves and the "scriptural" authors was in any way blurred or transgressed. Sedgwick deployed some of his most withering invective to criticize the professional science lecturer Andrew Ure, who after his election to the Geological Society revealed his true affinities by publishing *A New System of Geology* (1829): this claimed to "reconcile" the science with Genesis, but did so with gross inaccuracies on the scientific side. Yet the boundaries were not always so clear cut. The amateur fossil collectors George Young and John Bird (the former also a Presbyterian minister), in their *Geological Survey of the Yorkshire Coast* (1822), included such useful descriptions of local rock formations and fossils that their work could not be dismissed out of hand; yet their "young-Earth" interpretation of it all recalled Woodward's ideas more than a century earlier, and was literally incredible to any more experienced geologist.

In contrast to the members of the Geological Society, with their strong collective sense of identity and purpose, these other people who were thinking about the relation between geology and Genesis were strikingly diverse (they are now known to us primarily through those who published books and pamphlets, but their readers ranged far more widely). While some authors were ordained clergymen,

> By way of encouragement to my husband's labours, we have had
> the Bampton Lecturer holding forth in St Mary's against all mod-
> ern science (of which it need scarcely be said he is profoundly
> ignorant), but more particularly enlarging on the heresies and in-
> fidelities of geologists, denouncing all who assert that the world
> was not made in 6 days as obstinate unbelievers, &c. &c. . . . Alas!
> my poor husband – Could he be carried back a century, fire & fag-
> got would have been his fate, and I dare say our Bampton Lecturer
> would have thought it his duty to assist at such an 'Auto da Fé'.
> Perhaps I too might have come in for a broil as an agent in the
> propagation of heresies.

FIG. 7.1 Mary Buckland's report (1833)—in a letter to William Whewell, the outstanding English polymath of his generation and one of Sedgwick's Cambridge colleagues—on some prestigious lectures then being given in Oxford's university church by the Anglican clergyman Frederick No-lan (she exaggerated for rhetorical effect by dating the era of heretic-burning only one century in the past rather than three!). Whewell, Sedgwick, and her husband William Buckland were all practicing Christians and indeed "Reverend Professors," yet they often portrayed themselves and other geologists—however jokingly—as latter-day martyrs for science. Episodes such as this show how their encounter with literalistic or "scriptural" writers was no simple conflict between "Science" and "Religion."

many others were not. While some were Anglicans—and loyal to a church that in England, though not in Scotland, was closely tied to the state—others belonged to a variety of other Protestant bodies or were Roman Catholics. And their writings were far from being uniformly antagonistic towards the new science. At one end of a broad spectrum, some such as Bugg were indeed fiercely hostile to what they saw as the subversive tendencies of geology. What they particularly criticized were the geologists' claims about the allegedly lengthy pre-human history of the Earth, which, they claimed, flew in the face of the Creation narratives in Genesis and were bound to undermine confidence in the trustworthiness of the Bible as a whole. But many other authors were far more concerned to amplify or clarify the brief biblical narratives in the light of the geologists' surprising new discoveries. They argued, in a quite traditional style, that "God's Works" in the natural world could and should be used to complement "God's Word" in the Bible, and they welcomed the new ways in which these two sources of human knowledge might be amicably "harmonized" or "reconciled."

The sheer variety of these publications shows that any claim that there was a uniquely unambiguous "literal" interpretation of Genesis (or any other biblical text) was and is a misleading illusion. Most of their authors—like many of the leading geologists—did indeed regard the Bible as in some sense divinely inspired, but at the same time many were well aware of the problems of understanding ancient texts originally composed in ancient languages, which made any naïve literalism highly problematic.

The diversity of these works also stretched from the intensely scholarly to those aimed deliberately at a popular or juvenile readership. For example, Granville Penn, a civil servant in London but also a Classical scholar and philologist, published *A Comparative Estimate of the Mineral and Mosaical Geologies* (1822). This was a scholarly evaluation of what he regarded as rival accounts equally worthy of serious consideration, though he was convinced that Moses was a more reliable historian than the geologists. On a much simpler level, the anonymous *Conversations on Geology* (1828), written in fact by the science lecturer James Rennie, claimed to compare Penn's ideas even-handedly with those of Hutton and Werner, and with "the late discoveries of Professor Buckland" and other geologists, all under the guise of a well-educated mother enlightening her eagerly interested children.

Most revealing, however, is the frequency with which "scriptural" critics attacked the geologists' ideas for being *contrary to common sense*. The traditional public culture, particularly in countries as deeply Protestant as England and Scotland, was one in which people often considered themselves as much entitled as any self-appointed expert to judge things for themselves: as much in making sense of rocks and fossils as in making sense of biblical texts. In contrast, it could seem to them that geologists were claiming to have privileged access to deeper truths than those outside their charmed circle. To some, this recalled the exclusive claims of priests and other church authorities in an earlier age. So geologists in their own popular publications had to explain how their new ideas were based on what they had seen, especially *in the field*, which demanded a revision of what had previously seemed common sense, but which in principle was knowledge equally accessible to everyone.

However, all this fuss and bother about geology and Genesis remained almost exclusively confined to Britain and the United States

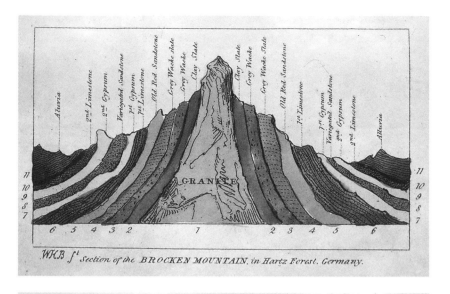

*WKB f<sup>t</sup> Section of the BROCKEN MOUNTAIN, in Hartz Forest, Germany.*

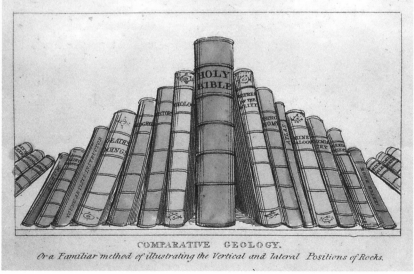

COMPARATIVE GEOLOGY.
*Or a Familiar method of illustrating the Vertical and lateral Positions of Rocks.*

**FIG. 7.2** A pair of illustrations from James Rennie's anonymous *Conversations on Geology* (1828), which was designed to make the science accessible and attractive to children. A diagrammatic slice or section through the Harz hills in northern Germany (above) shows a foundation of Primary granite flanked by a sequence of various Transition and Secondary rock formations and finally by Alluvium. In the matching picture (below), the Bible is the massive volume which—reassuringly!—is acting as a stable foundation for a stack of various other books (or perhaps their secular knowledge is propping it up?). This analogy applied the traditional notion of the "*book* of nature" to the new science of geology. It was not, in itself, tied either to the geologists' lengthy timescale for the Earth or to the "scriptural" writers' very brief one; on this point the fictional "Mrs R.," explaining the science to her dutifully attentive son and daughter, could sit comfortably on the fence.

(the latter still very close to Britain in cultural matters although proud of its political independence). Geologists in the rest of Europe tended to note with amused disdain how they, in contrast, could get on with their scientific work without having to defend their efforts against generally ignorant critics. And in practice their British colleagues also did so, as is clear from the scientific books they wrote for each other and the papers they read at their meetings and then published in a proliferating variety of scientific periodicals. Geologists were in fact united in an international scientific network, which was developing a broadly consensual view of the Earth's history. Those in Britain had good reason to worry sometimes about the public relations of their science, but they were right not to lose any sleep worrying about its validity (there is an obvious parallel here with the predicament of modern American scientists confronted by creationists). In the early 19th century the issue of geology and Genesis was little more than a storm in an Anglo-American tea-cup.

Far more significant in the long run was the far more widespread sense that the natural world was underlain by divine care, purpose, and design; or, in traditional language, by divine providence. However, mainstream Christian thinking had always regarded this kind of "natural theology" as subordinate or merely preparatory to the "revealed theology" based on what was believed to be divine self-disclosure through the events of human history. In early 19th-century Britain, for example, William Paley's classic *Natural Theology* (1802) had famously offered an eloquent formulation of the traditional "argument from design," the claim that the apparent designfulness of the natural world points to the providential purposes of a divine Designer (the argument now revived by creationists as "Intelligent Design"). But this work was treated merely as a supplement, or at most a complement, to Paley's earlier *Evidences of Christianity* (1794), which had set out the far more important *historical* grounds for specifically Christian faith. Nonetheless, natural theology was widely considered a valuable source of intellectual support for religious belief in general, not least because its persuasive appeal united people across the widest range of theological and ecclesiastical loyalties, from Unitarians and Quakers to Roman Catholics.

In Britain the "Bridgewater Treatises," a distinguished series of books funded posthumously by a wealthy aristocratic savant (and ordained clergyman), were intended to update Paley's argument in the light of the spectacular advances then being made in many of the natural sciences. Whewell, for example, was commissioned to write on astronomy, and Buckland on geology. Buckland's *Geology and Mineralogy* (1836) duly offered other geologists and the general reading public an impressive and authoritative review of his science. At the same time he transcended Paley's argument by giving it the new dimension of deep history. For example, his analysis of the anatomy of the long-extinct trilobites—some of the oldest fossils then known—claimed in effect that animals had *always* been well designed, and therefore well adapted to follow specific ways of life, even back in the remotest depths of the past. (This kind of analysis of fossils, far from having "Retarded the Progress of Science" on account of its religious roots, continues to underpin the functional reconstruction of fossils by

FIG. 7.3 "The country of the Iguanodon": the frontispiece of Gideon Mantell's *Wonders of Geology* (1838). It depicted the huge monsters whose fragmentary fossil remains he had first discovered, set in a tropical scene of Sussex, in the south of England, as it was in the remote past. Like other authors of popular works on geology, Mantell stressed the romantic "wonder" of the deep history of the Earth and its life, which this exciting but unfamiliar science was revealing. This "scene from deep time" was drawn for him by John Martin, an artist already famous for painting similarly melodramatic pictures of scenes from *human* history, both sacred and secular, such as the Fall of Babylon and the eruption of Vesuvius that had destroyed Pompeii.

modern scientists, who talk quite casually about an organism's *design* without doubting that the cause of the design has been its evolution through natural selection; my own palaeontological research before I became a historian was of just this kind.)

Closely related to this sense of the providential designfulness of the natural world was a sense of wonder at the romance of the vanished deep past that the geologists' research was disclosing. So, for example, Mantell—who had discovered the *Iguanodon*, the first of the fossil reptiles to be classed later as a dinosaur—exploited a profitable vein of popular science by describing the *Wonders of Geology* (1838). The sheer scale and unanticipated strangeness of the Earth's long history was often treated as welcome new evidence for the grandeur of God's creation. Far from geology being in intrinsic conflict with religious faith, the science was widely regarded in the early 19th century as its ally and supporter.

## A DISCONCERTING OUTSIDER

The new science of geology, then, offered a dramatically persuasive reconstruction of some of the main outlines of the history of the Earth and its life. In the account given up to this point, however, the unmentioned elephant in the room has been Charles Lyell. Of all the geologists of his generation he is now one of the best known, at least by name, to his modern successors. He is often portrayed as a "Founding Father" of geology and as a heroic figure who, with only Hutton as forerunner, first demonstrated the Earth's vast timescale, vanquished the reactionary forces of Religion, and paved the way for the evolutionary theory of his younger friend Charles Darwin. Lyell was indeed one of the finest geologists of his time, and his work had a lasting impact on his science, but he deserves to be evaluated in a more historical way than this crude hagiography allows.

As a young man Lyell was recognized as a rising star in the world of geology. While a student at Oxford, he attended Buckland's lectures and was greatly impressed by them. He then trained as a lawyer in London, and also joined and became an active member of the Geological Society. In articles for the influential *Quarterly Review*, he outlined some of the geologists' latest discoveries and ideas for its intelligent British readers, in a way that shows that he was in the mainstream of his science, expounding the directional kind of Earth history that his older colleagues were busy reconstructing. However, Lyell had also read Playfair and was impressed by his arguments for the power of "actual causes" or present geological processes. He became convinced that Cuvier—whom he greatly admired—had been too quick to dismiss them as inadequate to explain mass extinction events and other apparently sudden "revolutions" in the deep past. While Lyell was on a visit to Paris, the geologist Constant Prévost took him into the field to see the famous Parisian sequence of Tertiary formations, and convinced him that the freshwater ones had been formed in "the former world" under just the same conditions as those of the present day. Lyell then confirmed this for himself by finding closely similar sediments of "the present world" in a recently drained lake near his family home in Scotland. When his near contemporary George Poulett Scrope published a detailed description of the famous extinct volcanoes of central France—reviving Desmarest's interpretation of them as a record of a very long and complex sequence of eruptions—Lyell agreed with him emphatically: geologists were failing to appreciate fully the impli-

The periods which to our narrow apprehension, and compared with our ephemeral existence, appear of incalculable duration, are in all probability but trifles in the calendar of nature. It is Geology that, above all other sciences, makes us acquainted with this important, though humiliating fact. Every step we take in its pursuit forces us to make almost unlimited drafts upon antiquity. The leading idea which is present in all our researches, and which accompanies every fresh observation, the sound of which to the student of Nature seems continually echoed from every part of her works, is —

Time! − Time! − Time!

FIG. 7.4  A famous quotation from Scrope's *Geology of Central France* (1827). It expressed his belief that although geologists claimed to accept that the Earth's timescale was almost inconceivably lengthy, in practice they were failing to appreciate the implications of this for explaining geological features in terms of observable present processes. Scrope was a member of Parliament and wrote extensively on political economy, including currency reform; the "drafts" of deep time were, metaphorically, to be withdrawn from a bottomless bank account. This idea of the unlimited explanatory power of deep time was eagerly adopted by Charles Lyell and became a leading theme in all his work.

cations of the vast timescale that they all professed to accept. Given enough time, ordinary physical processes could produce huge physical effects.

Scrope also convinced him that at least in this classic region there was no clear sign of any recent Deluge. Lyell began to suspect that Buckland's diluvial theory might be altogether mistaken. He was particularly sceptical about his mentor's identification of the alleged geological event with the biblical Flood. This was reinforced by his growing antipathy not toward religious beliefs as such, but toward the political and cultural power of the Church of England, and above all its monopoly of higher education in England, as embodied in Buckland's Oxford. Lyell planned to reject not only the alleged evidence for the biblical Flood and geological Deluge, but also the wider arguments for the reality of still earlier "catastrophes" of any kind. Conversely he would persuade geologists to recognize the sheer power of "actual causes" or present geological processes—when played out on an almost inconceivably vast scale of deep time—to explain the same evidence more satisfactorily.

On his first major geological tour in mainland Europe, Lyell made the extinct French volcanoes his first priority, and on the spot he was fully persuaded by Scrope's interpretation of them. Further fieldwork all down the length of Italy, including the active volcano Vesuvius, convinced him that present geological processes were much more powerful than most geologists realized, and that a fuller knowledge of them would be the key to a true understanding of the Earth's history. On the island of Sicily, the huge active volcano Etna, which had clearly accumulated—one lava flow after another—on top of a thick pile of Tertiary formations, confirmed his sense of the vastness of the total time that had been available for present processes to have produced even the largest effects. A high mountain range, for example, could have been heaved up not by a single mega-earthquake, as Élie de Beaumont and others suggested, but very slowly by a lengthy sequence of ordinary earthquakes, none more powerful than those recorded in human history.

Lyell determined to match the then current campaign for political reform in Britain—a first step towards expanding the right to vote—with a comparable reform of the science of geology. While still in the field he told Murchison that the book he was planning to write would be based on two fundamental "*principles of reasoning*" in geology. The first was that "*no causes whatever* have, from the earliest time to which we can look back, ever acted, but those *now* acting" (the emphases were his). This made the principle of actualism much more rigorous than was usual among geologists, by ruling out the possibility that some past processes might no longer be active or might not yet have been seen in action in the present world. Lyell's second principle was even more rigorous: it was "that they [present processes] never acted with different degrees of energy from that which they now exert." This was still more questionable: it implied, for example, that, whatever the physical processes that cause tsunamis, they could never have acted in the deep past with much greater intensity, generating a mega-tsunami unparalleled in recorded human history. Lyell believed that this principle, if applied as consistently as he intended, would necessarily lead him to reject his contemporaries' conception of the Earth's history and to adopt instead something like Hutton's earlier steady-state system. It would be, as he put it, a system based on "absolute uniformity": no overall directional trends, no exceptional catastrophes.

After he returned home, Lyell wrote a massive three-volume *Principles of Geology* (1830–33), designed to establish this kind of system. He would try to explain *all* the traces of the deep past in terms of "causes now in operation," such as erosion and deposition, crustal elevation and subsidence, volcanoes and earthquakes, the physical impacts of plants and animals, and so on. His first two volumes were an exhaustive inventory of the effects that these processes had had *within recorded human history*. Lyell borrowed much of his material from a huge recent compilation by the German civil servant and historian Karl von Hoff (Lyell learned German in order to read it). He used this evidence to argue that the underlying processes were in dynamic equilibrium: erosion balanced by deposition, crustal uplift balanced by subsidence, the formation of new species balanced by the extinction of old ones, and so on. Changes were cyclic, and in the long run the Earth was maintained in a steady state. The frontispiece that acted as a visual summary of his whole argument displayed not some striking geological feature but—surprisingly—a Classical ruin. But this was a site that displayed the Earth's steady state in miniature, within the span of recorded human history.

Lyell claimed that his huge inventory of present processes provided

**FIG. 7.5** The frontispiece of the first volume (1830) of Lyell's *Principles of Geology*, which acted as a visual summary of the whole work. The remaining columns of the ruined Temple of Serapis near Naples (later re-interpreted as a market building) were marked by a zone in which the stone had been bored by marine molluscs, obviously at a time when the level of the almost tideless Mediterranean Sea was higher. Lyell used the ruin as evidence that within the two millennia since Roman times the land in this earthquake-prone volcanic region had subsided and later risen again, almost to its earlier level (the marble pavement was still just flooded by sea water), yet gently enough for the columns to have remained upright throughout. This illustrated in miniature Lyell's interpretation of the Earth as being in a steady state of dynamic equilibrium, without violent disruptions; it also embodied the actualistic method of using the span of human history as the key to the deeper past. The engraving was copied from one published by a Neapolitan antiquarian, but Lyell had seen the ruin for himself during his Italian travels.

the essential "alphabet and grammar" of geology. In his third and culminating volume he argued that this enabled geologists to decipher
nature's "language" in which the Earth's own historical documents had
been recorded, and so to reconstruct its deep history. Jean Champol-
lion's recent and much celebrated deciphering of the ancient Egyptian
hieroglyphs made this a vivid metaphor, exploiting yet again the analogy between geology and human history. Taking his cue from Cuvier and others going back to Steno, Lyell reconstructed the Earth's
history retrospectively, moving from the observable present into the
unobservable and increasingly unfamiliar past, and focusing on the
nearest part—the Tertiary era—to exemplify that strategy. Some of
the best evidence for this era came from its abundant fossil shells. Lyell adopted Brocchi's model of piecemeal change among its species
of molluscs, making an explicit analogy with the population censuses
that had begun to be taken every ten years in Britain. The Tertiary
formations scattered around Europe (in the "basins" of Paris, London,
and elsewhere) could then be arranged in *chronological* order—and
quantitatively, although on a scale not calibrated in years—by counting among their fossils those species known to be *extant* or still alive
and those unknown alive and probably extinct: the higher the per-
centage of extant species the more recent the formation (for identifying the hundreds of species he relied on the expertise of a Parisian
naturalist). Whewell then supplied Lyell with suitably Classical tags
to name the periods *within* the Tertiary era from which these popula-
tions of species happened to have been preserved. They ranged from
*Eocene* ("dawn of recent [i.e., extant] species") through *Miocene* ("less
recent") to *Pliocene* ("fully recent"; the names remain in use by modern geologists, along with others inserted later).

Lyell noted the almost total disparity between the fossils of the Eocene, the oldest of these Tertiaries, and the youngest of the underlying
Secondaries (the Chalk at Maastricht). Revealingly, he interpreted
this as an unpreserved gap in the fossil record as lengthy as the *entire
Tertiary era* that followed. This inference—as startling to geologists at
the time as it should be to their modern successors—followed logi-
cally from Lyell's claim that the rate of change had been statistically
uniform throughout: a steady rate of extinctions balanced by a steady
rate of new species. It showed his principle of "absolute uniformity"
in action. It also implied that the fossil record, far from being a fairly

complete inventory of the history of life, as most other geologists believed, must in fact be extremely imperfect and fragmentary.

Finally, after a brief survey of the Secondary formations, just to suggest how the same kind of analysis might in the future be extended still further back in time, Lyell's *Principles* concluded with a summary of his model of the Earth's history: there had been steady or cyclic change from as far back as any record survived, with no overall directional trends and no exceptional "revolutions" or "catastrophes" on the way.

## CATASTROPHE VERSUS UNIFORMITY

Lyell's model was in effect an updating of Hutton's steady-state *Theory of the Earth*. It was incompatible with the directional model of a slowly cooling Earth and a broadly "progressive" history of life, which was adopted by almost all other geologists: this included the many, such as Prévost and Scrope, who fully agreed with Lyell about the vast scale of time and the value of using present processes as the best key to the deep past. So Lyell had to work hard to explain *away* their strong conviction that the Earth and its life had *not* always been in more or less the same overall state. Lyell was convinced that the fossil record was extremely fragmentary and imperfect, because the chance of any animal or plant being preserved had always been very low. So he saw no problem about inferring a huge unpreserved gap between the Tertiary formations and the Secondaries. Similarly, other geologists regarded the small and very rare marsupial mammals in Jurassic strata as further evidence for the directional character of the whole history of quadruped life: they were far earlier than any other mammals and arguably more primitive. But to Lyell the finding of *any* mammal in such strata showed that a full range of mammals might already have been in existence even further back in the Earth's history, although none happened to have been preserved. And where the fossil record petered out altogether, in the very ancient formations that Sedgwick soon afterwards named "Cambrian," Lyell argued that it was because still earlier strata (later named pre-Cambrian or Precambrian) had been so thoroughly altered by intense heat in the depths of the Earth that all traces of their fossils had been destroyed: he called such rocks *metamorphic* ("transformed"). But what his contemporaries found

**FIG. 7.6** De la Beche's caricature (1830) ridiculing Lyell's suggestion that the celebrated Jurassic reptiles—or at least, animals much like them—might *return* in the distant post-human future, when suitable environmental conditions recurred in the course of the Earth's vast cycles of change. "Professor Ichthyosaurus" is lecturing to an audience of other reptiles, interpreting a fossilized human skull as a trace of long-extinct animals of a lower class than themselves. In the opinion of other geologists, Lyell's cyclic or steady-state interpretation of the Earth's long-term history, which was at the heart of his massive *Principles of Geology*, was utterly implausible.

most implausible of all was that Lyell suggested in all seriousness that the age of giant reptiles might eventually *return*, as the global physical conditions of the Jurassic period, in which they had flourished in the deep past, were repeated in the distant future, in the course of the vast cycles of the Earth's steady-state history.

In the face of this radically contrarian interpretation of the Earth's history—which seriously disrupted the comfortable consensus among other geologists—Whewell suggested that geologists were now divided into two opposing sects: he was alluding to the vehement religious controversies agitating British public life at this time. Lyell's insistence on the "absolute uniformity" of geological processes identified him as a *"Uniformitarian,"* though Whewell pointed out that this was a very exclusive sect indeed (the young geologist Charles Darwin, soon to return from his voyage around the world on the *Beagle*, became one of Lyell's very few significant followers). Lyell's critics formed a much more populous denomination, which Whewell dubbed "*Catastrophist*": catastrophists not only championed a directional kind of history for the Earth and its life but also one that seemed to have been punctuated occasionally by the sudden natural events that they called "revolutions" or "catastrophes." Contrary to later misconceptions about this argument, *both* geological sects adhered to the doctrine now termed actualist, namely that the present is generally the best key to the past; they differed only in their view of the *adequacy* of "actual" or present processes, *at their present intensity*, to account for everything in the deep past. Likewise the magnitude of the Earth's total timescale was not at issue, though Lyell often claimed for his own rhetorical purposes that it was, and that his critics were only invoking catastrophes because they were squeezed for time. Conybeare, for example, protested that he and other catastrophists would be happy to allow "quadrillions of years" (far in excess of modern estimates of a few billions) if the evidence turned out to require it; but he pointed out that no amount of time, on its own, would or could eliminate the evidence for a directional history of the Earth.

Sedgwick complained publicly that Lyell in his *Principles* used too much "the language of an advocate"—which as a qualified barrister was just what Lyell was!—but Sedgwick himself was an eloquent preacher, equally accomplished in the arts of rhetoric. Both sides quite properly deployed their evidence in the most persuasive way possible, making as strong a case as they could. But in fact this vigorous but generally amicable debate eventually subsided into something of a draw. Other geologists learned from Lyell a better appreciation of the power of present processes, acting over vast spans of deep time, and they conceded that in some cases apparently catastrophic changes might indeed have happened slowly and gradually. But they adamantly declined to follow Lyell in interpreting the Earth as a cyclic or steady-state system. On the contrary, their research seemed to turn up more and more evidence for its directional and *historical* character, whether or not this was attributed to a steady cooling process or to some other cause. And they remained highly sceptical about Lyell's attempt to explain away *all* the evidence for occasional natural catastrophes in the deep past. Anyway this debate was largely confined to geologists in Britain. Prévost, Lyell's leading ally in France, had planned to translate his *Principles* but was distracted by the political events of the July Revolution and never did so. What was later translated into French and other languages was Lyell's valuable inventory of the directly observed effects of geological processes within recorded human history. This encouraged geologists everywhere to try to interpret the past history of the Earth *wherever possible* in terms of ordinary processes, and not to rush prematurely into invoking unusual or exceptional events unless the evidence for them was overwhelming. On the other hand, Lyell's steady-state history of the Earth, which had been split off into a separate work (*Elements of Geology*, 1838), was never given as much attention even in the Anglophone world, and still less outside it.

Meanwhile, in Britain, Lyell's eloquent prose made his *Principles* almost as accessible to the educated public as it was to his fellow geologists. What most impressed the public was his persuasive evidence for a vast timescale for the Earth and his scornful dismissal of the "scriptural" writers as ignorant and worthless. But these were of course two issues on which other geologists, including the many who were religious, fully agreed with him. Yet Lyell's skillful lawyer's rhetoric gave the public the impression that to be truly scientific in geology was to be a Lyellian or uniformitarian, and that the catastrophist geologists were little better than the "scriptural" writers. His geological critics naturally protested that this was grossly unfair.

Paradoxically, the most recalcitrant case of an apparent catastrophe lay not in the deep past but close to the present. What had happened in the geologically recent past, or around the dawn of human history, was more obscure than what had happened much further back in the past. The puzzling Superficial deposits that had been termed *diluvium*, particularly the distinctive till or boulder clay, were unlike anything known to be forming at present, or anything known to have formed in the more distant past. The alleged geological Deluge had, not unnaturally, been identified at first with the biblical Flood, as the only recorded event in human history of possibly adequate magnitude. But further fieldwork on the diluvium in various parts of Europe had soon shown that these deposits, or most of them, almost certainly dated from much too far in the past to be equated with the biblical Flood, and indeed that there might have been more than one such diluvial episode. Sedgwick called his own conversion on this point a "recantation"—the usual jokey allusion to much earlier religious

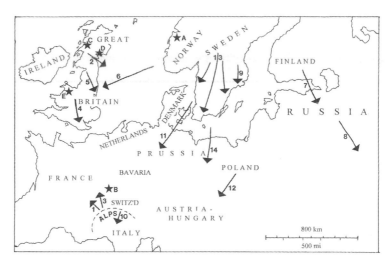

**FIG. 7.7** A map of the "diluvial" currents that were traced across Europe in the early 19th century, on the evidence of erratic blocks, scratched bedrock and other features. Each case was described by a specific geologist: 1, by Horace-Bénédict de Saussure; 2, by Hall, 3, by von Buch, 4, 5, and 6, by Buckland; 7 and 8, by Gregor Razumovsky and William Fox-Strangways; 9, by Alexandre Brongniart; 10, by De la Beche; 11, by Johann Hausmann; 12, by Georg Pusch; 13 and 14, by Nils Sefström (as the names suggest, this research was highly international. All these cases were later (from the 1840s) re-interpreted as the traces of vast *ice-sheets* during a Pleistocene "*Ice Age*." Crucial to that dramatic change were the areas with traces of much smaller and local *valley glaciers*, shown here by stars: (A, Norway; B, the Vosges; C and D, the Scottish Highlands; E, North Wales).

heresy-hunting—but in reality it was a quite minor alteration. Even Buckland, who had so strongly identified the Deluge with the Flood, later changed his mind without anguish or embarrassment. Lyell thought that this should have forced his critics to abandon *any* connection between geology and the biblical event; but of course it was still open to them to treat the Flood story as a faint record of a probably local event in early human history—as the best biblical scholarship of the time suggested—while continuing to insist that the much earlier geological Deluge needed to be explained and not merely explained away. As mentioned earlier, the diluvial theory, far from having its plausibility undermined, went from strength to strength, as erratic blocks, scratched bedrock, and other features were traced ever more widely across Europe and North America.

Lyell tried hard to explain away all the alleged traces of a geological Deluge, by invoking what he called his new theory of climate. He pointed out that local climates depend not just on latitude but also on the disposition of continental landmasses and ocean currents. Britain, for example, enjoys a temperate climate in sharp contrast to the frigid conditions in Labrador across the North Atlantic, although they are on the same latitude. If then the region that now forms Europe was formerly deprived of the warmth of the Gulf Stream, icebergs from the Arctic regions might have drifted much further south than they do at present. If at the same time the sea level was higher, they could have dropped their cargo of erratic blocks as they melted (in modern terms, as *dropstones*) all across what is now low-lying northern Europe. Lyell's interpretation was quite plausible for these northern erratics, but it was much more awkward to apply to the Alpine ones, which were often found at much higher altitudes. And it did not explain satisfactorily the widespread occurrence of scratched bedrock, or the equally widespread and peculiar deposits of till or boulder clay. Nonetheless, despite these drawbacks, Lyell reinterpreted all erratics—the strongest evidence for a geological Deluge—as dropstones from drifting icebergs, and the diluvium as a whole as what he called "*drift*" deposits. In this way his *drift theory* neatly eliminated any hint of a "catastrophe" in the geologically recent past, and on a global level it preserved an overall steady-state climatic "uniformity."

Lyell's drift theory gained some support from the discovery that some of the youngest Tertiary deposits in Britain, which he had named "*Newer Pliocene*," contained fossil shells of species that now live only in much colder and more northerly waters. He re-named this period "*Pleistocene*" ("most recent"; his earlier "Older Pliocene" was then redefined as just "Pliocene"). This apparently minor name change subtly converted the alleged diluvial period into an ordinary part of the Tertiary era, implicitly undermining the claim that in the geologically recent past there had been any event of a radically different and "catastrophic" character. However, in the eyes of his critics there was much that Lyell's drift theory failed to explain satisfactorily, and the diluvial theory continued to have the support of most geologists.

## THE GREAT "ICE AGE"

An alternative explanation of these puzzling features, which ultimately proved far more satisfactory than either the diluvial theory or

Lyell's drift theory, emerged from an unexpected direction. The Swiss civil engineer Ignace Venetz had reported that people living in Alpine valleys knew that the glaciers had fluctuated in size and extent, even within historical times. This was shown in particular by the *moraines*, or stony ridges along the sides and at the snouts of glaciers, where the ice melts and drops the rock debris it has picked up. Venetz had described similar moraines much further down the valleys and much higher up on their flanks (often concealed by forest); he had argued that the Alpine glaciers must have been far larger at "an epoch *that is lost in the night of time.*" However, other Swiss savants—while meeting, appropriately, high up on the Great St Bernard pass—ignored this claim or dismissed it as a wild conjecture. But one of these sceptics, the geologist Jean de Charpentier, later became convinced that it alone could account for what he knew about his own part of the Alps. There were moraines high up on the sides of the upper Rhône valley, and some huge erratics were perched there; and up to that level there were scratched or polished surfaces of bedrock. Crucially, he knew that similar rock surfaces could be seen underneath the ice of existing glaciers much further up the valley, having obviously been scratched by the boulders frozen into the moving ice (like coarse sandpaper scratching a surface of wood). Charpentier traced the moraines and scratched bedrock all across the region, and claimed sensationally that a huge glacier must have formerly filled the entire upper Rhône valley, spreading out right across the low-lying Swiss Plain beyond the Alps and even pushing up on to the Jura hills to the north, where it left behind some famous huge erratics. In this historical reconstruction, even the largest of the existing Alpine glaciers were reduced to tiny shrunken remnants of a once vastly more extensive "*mega-glacier.*"

This implied that snowfall on the Alps, eventually compacted into glacial ice, must have been much greater in the geologically recent past. What could have caused this? Like most other geologists Charpentier could not imagine that the *global* climate might have been much colder: all the evidence for the long-term cooling of the Earth implied that it would have been, if anything, a little warmer. However, if at that time the Alps had been as high as the present Andes or Himalaya, they might have generated mega-glaciers even in an overall global climate much like the present. But this would have required a huge elevation of the Alps—Charpentier clearly had Élie de Beau-

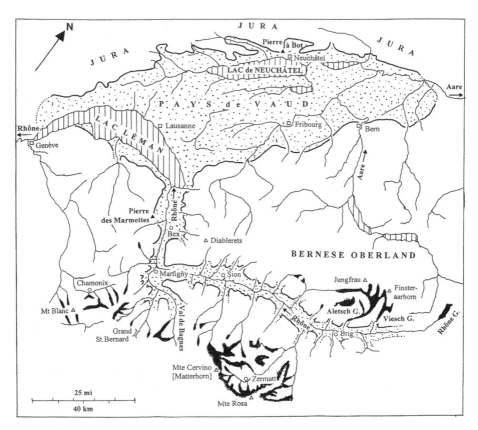

**FIG. 7.8** Jean de Charpentier's map of western Switzerland (published 1841), showing his reconstruction of a former *glacier-monstre* or "mega-glacier"(stippled) filling the whole of the upper Rhône valley, spreading out across the site of the present Lac Léman (Lake of Geneva) and over the low-lying Pays de Vaud, and pushing up on to the flanks of the Jura hills. In effect, this claimed to replace von Buch's earlier interpretation of much the same features in terms of a diluvial current or mega-tsunami. Charpentier's reconstruction was based on his detailed mapping of the whole "erratic field" and especially the moraines on the sides of the valleys (flanking the stippled area). The existing glaciers, including the Aletsch, the longest in the Alps, are shown (in black) as relatively tiny remnants of the vanished mega-glacier. The positions of two famous giant erratics are also marked, one close to Charpentier's home (*Pierre des Marmettes*), the other on the Jura above Neuchâtel (*Pierre à Bot*). His large and detailed map has here been redrawn in modern style (and very greatly reduced in size) to clarify his interpretation.

mont's periodic "epochs of elevation" in mind—followed by their subsidence back to their present height, all within a geologically short span of time. Other geologists found this explanation unconvincing. Furthermore, while it accounted for the Alpine erratics far better than von Buch's earlier idea of some kind of mega-tsunami, it could not apply to the otherwise similar erratics in the north of Europe and in North America: these were not close to any mountain range, let alone one high enough to account for their vast spread. So Charpentier's theory of a huge former extension of the Rhône glacier was treated by other geologists with cautious scepticism.

However, it soon became the catalyst for an even more sensational theory, which was proposed by Agassiz when in 1837 the annual meeting of Swiss naturalists was held in his home city of Neuchâtel. Agassiz was already well known and respected for his work on fossil fish, but here he launched out into a quite different field in which he had no previous experience. He suggested that in the geologically recent past the Earth had been in the grip of a drastic "*Ice Age*." He argued that this had been so cold that a *static* sheet of snow or ice had covered the entire Northern Hemisphere, at least as far south as the Atlas mountains in North Africa; he probably thought—as he certainly did later— that in fact it extended to the tropics, creating what has since (in a different context) been termed a "*Snowball Earth*." During this extremely severe Ice Age the Alps had been heaved up—like Charpentier he evidently had Élie de Beaumont's theory in mind—creating an icy slope down which erratic blocks of Alpine rock had slid all the way to the Jura hills (at the foot of which the naturalists were meeting). Only later, as the global climate emerged from this Ice Age, had the static sheet of snow and ice melted away, leaving the present slow-moving glaciers as its puny remnants. Agassiz minimized his debt to Charpentier and claimed that his own theory was quite different, as indeed it was. Unlike his rival, he did not think erratics had been transported by the moving ice of any glacier, but simply by sliding down a sloping surface of static ice.

Agassiz also ingeniously integrated his idea of a brief but extreme Ice Age with the slow overall cooling of the Earth that he along with most geologists (other than Lyell) found highly plausible. He conjectured that the Earth had not cooled gradually but in a series of steps, each providing a period of stable environments to which a specific set of animals and plants could have been well adapted. These stable periods had been separated by sudden temporary plunges in global temperature, the cause of which he left conveniently vague. This could account for the repeated episodes of mass extinction that had given each successive period its distinctive fauna and flora in the fossil rec-

**FIG. 7.9** Agassiz's picture of a surface of scratched bedrock at the foot of the Jura hills near Neuchâtel, published in his *Studies on Glaciers* (1840) to illustrate his theory of a drastic Ice Age in the geologically recent past. In fact such evidence fitted *Charpentier's* theory—of a former mega-glacier extending all the way from the Alps and carrying erratic blocks that did the scratching while embedded in its ice—better than it matched Agassiz's own theory that the erratics had merely slid down a sloping surface of static ice. Along with erratic blocks, and till or boulder clay, scratched rock surfaces such as this became crucial evidence for former glaciers or even ice-sheets across vast areas of northern Europe and North America.

ord. Superimposed on the Earth's long-term cooling trend, only the most recent of these sudden drops in temperature had plunged low enough to cause an Ice Age. A very long-term trend had thus generated a unique event in the very recent past.

This was speculative theorizing on a heroic scale. Not surprisingly, more senior and sober geologists such as von Buch, who was present at the Neuchâtel meeting, were highly sceptical; others suggested bluntly that Agassiz should stick to his fossil fish. But some were sufficiently intrigued to look again at their home areas. For example, geologists living near the hills of the Vosges (in Alsace, to the north of the Jura hills and still further from the Alps) found abundant evidence of small former glaciers in the deep valleys, now close to the warm hillsides that produce fine Alsatian wines. There was no sign that the Vosges had been heaved up and then subsided again in the recent past, so Charpentier's explanation for his Alpine mega-glacier could hardly apply there. The evidence seemed to call instead for a genuinely cold period of at least regional extent.

Agassiz himself visited Britain, primarily to study collections of fossil fish. But he also explained his Ice Age theory at a meeting of British "men of science" (the gendered but factually accurate term often used in Britain at this time). Then he was taken by Buckland on a tour of the Scottish Highlands, where they too recognized clear and widespread evidence of vanished valley glaciers. They also believed they saw traces of much more widespread ice across much of the lowlands of Scotland: sensationally, Agassiz claimed, for example, that the Old Town or city center of Edinburgh topped by its castle had once been a rocky island (in modern terms, a *nunatak*) surrounded by a sheet of glacial ice. Even Lyell was convinced at first, when in Buckland's company he saw similar evidence of ice sheets near his family home at the southern edge of the Highlands. But on reflection he thought this a glacier too far, and he retreated to a more moderate theoretical position. Valley glaciers in upland regions, like those now at higher altitudes in the Alps, seemed credible to him, but not extensive ice sheets across lowland areas. Small *local* glaciers were compatible with his theory of climate, as fluctuating in line with regional geographical changes; but lowland ice sheets would have required a *global* period of intense cold that departed too far from strict uniformity to be allowable on his principles. Even this much of an Ice Age looked unnervingly like what his critics called a catastrophe.

In fact, other geologists, including leading catastrophists, also found the "Bucklando-Agassizean Universal Glacier"—as Conybeare jokingly called it—literally incredible. Yet further cases of vanished valley glaciers in other hilly areas such as north Wales (where they were recognized by Darwin, to his own surprise) made some kind of "*glacial period*," equated more or less with Lyell's Pleistocene period, increasingly plausible, at least for temperate regions such as northern Europe. Agassiz reprinted his much more dramatic theory of a Snow ball Earth at the end of his *Studies on Glaciers* (*Études sur les Glaciers*, 1840). But this was mostly a description of the Alpine glaciers, so it was valued mainly for making *present* glacial action better known. In the years that followed, geological opinion congealed around a synthesis that combined a moderate "*glacial theory*" with Lyell's drift theory.

Most geologists agreed that the Earth, or at least its Northern Hemisphere, had passed recently through a period of much colder climate, but an "Ice Age" much less extreme than Agassiz's. In middling latitudes it had been cold enough to generate small valley glaciers. Those that were close to a coast would have produced icebergs that drifted away with their cargo of rocks, which were dropped off over much wider areas as the icebergs melted. Where the land had subsequently risen (or the sea level had fallen), these dropstones would be left as erratic blocks scattered across low-lying land. Modern parallels for this interpretation were drawn with Spitsbergen (Svalbard) in the north and South Georgia in the south, where valley glaciers extend down to sea level and, as they melt, calve icebergs directly into the sea. All the earlier "diluvial" evidence for some kind of mega-tsunami could now be deployed without loss, as it were, by being re-interpreted in terms of a period of much colder global climate. The diluvial theory was easily transformed into a glacial theory.

This glacial theory disturbed the earlier geological consensus, because it was utterly unexpected: not only to the majority of geologists who believed the Earth had been cooling slowly and steadily throughout its long history, but also to Lyell, who thought that it had always maintained a fairly equable steady state. Few if any geologists had anticipated finding that the Earth had passed recently through a geologically brief period of quite drastic cold, *and then warmed up again.* If any of them felt vindicated by the glacial theory, it was the catastrophists. For it reinforced the deeper sense, which catastrophists had always emphasized, that the past history of the Earth had been radically *contingent* and was therefore unpredictable even in retrospect. By this time the analogy with human history was so much taken for granted that it was no longer often used explicitly by geologists (except in what they wrote for popular consumption), but in fact this is just what the glacial theory confirmed. To reconstruct the Earth's history, a geologist needed to think like a historian, and in retrospect to expect the unexpected. The next chapter traces the further implications of this.

# 8

# Human History in Nature's History

In the second half of the 19th century the reconstruction of the Earth's history, reaching all the way back to the apparent start of the fossil record in the ancient Cambrian rocks, continued—paradoxically—to be most enigmatic at its near end, for the relatively recent past. What had happened to the Earth and its life in the puzzling interval between the time of the younger Tertiary strata and the present day? To fill that gap, the utterly unexpected idea of a drastic Ice Age was as "catastrophic" in character as the geological Deluge that it replaced. Agassiz's startling notion of a Snowball Earth had soon been discounted by other geologists, though he himself continued to argue for it (after he left Switzerland and settled in the United States as a professor at Harvard, he claimed to find traces of glacial action in tropical Brazil). But even the moderate kind of glaciation that was taken by most other geologists to have characterized Lyell's "Pleistocene" period—featuring widespread valley glaciers and seas strewn with icebergs—was still unexpected enough to accentuate the contingent character of the Earth's entire history.

Around the middle of the century, however, geologists' understanding of the effects of ice as a Lyellian "cause now in

operation" was dramatically enlarged as a result of polar exploration. In the north, expeditions in search of a commercial and stra-
tegic Northwest Passage around the top of North America, and the
voyages of whalers in search of ever-shrinking stocks, confirmed
that the hazardous icebergs drifting in the North Atlantic shipping
lanes were being calved from glaciers in Greenland. A scientific expedition then discovered that these were no mere valley glaciers, but
were spilling down from a huge ice sheet covering most of the interior
of that vast landmass. In the south, expeditions trying to locate the
south magnetic pole—to help understand the Earth's magnetic field,
on which the use of the magnetic compass for worldwide navigation
depended—penetrated the vast tracts of the southern ocean strewn
with similar icebergs. There they discovered a huge Antarctic continent, long suspected but never previously sighted, almost totally covered by an ice sheet even greater than Greenland's.

These discoveries suggested a more radical reconstruction of the
Pleistocene glacial period. Greenland and Antarctica, rather than the
Alps, became the relevant analogues in the modern world for what
northern Europe and northern North America might have been like
in the geologically recent past. Small vanished valley glaciers, for example those in Scotland, might have been just the last remnants of far
larger ice sheets that had previously spread out across lowland areas.
Such ice sheets had been suggested earlier, to account for the erratics
in Scandinavia and Germany; but like Agassiz's similar interpretation
of the Scottish evidence, the idea had seemed too wild to be taken
seriously. In 1875, however, the Swedish geologist Otto Torell—who
knew the Arctic glaciers and ice caps in Spitzbergen at first hand—
convinced leading members of the German Geological Society in Berlin that a huge thick ice sheet like Greenland's had not only covered
the whole of Scandinavia but also extended southwards across the site
of the Baltic Sea and over the north German plain, carrying with it
all the erratics that had previously been attributed either to a mega-
tsunami or to drifting icebergs. This was eventually accepted by geologists generally, and by the end of the century a similar but even more
extensive former ice sheet had been mapped across northern North
America.

The Ice Age, like the geological Deluge that it replaced, had at first
been assumed to be a unique event in the quite recent history of the

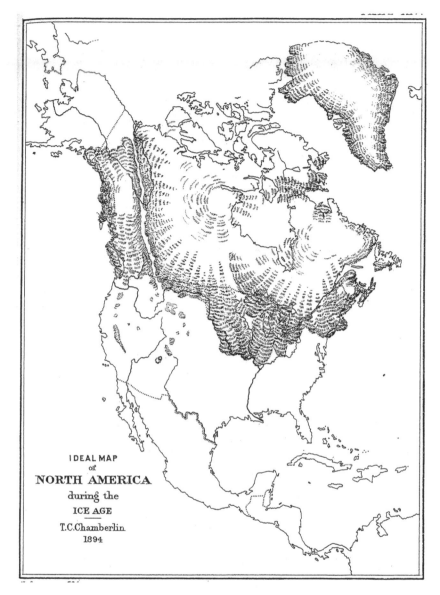

**FIG. 8.1** Thomas Chamberlin's "Ideal Map of North America during the Ice Age," as published in 1894 in James Geikie's *The Great Ice Age*. This reconstruction of the vast ice sheets, which at their maximum had covered the north of the continent, was based on the mapping of moraines and other glacial features by many American and Canadian geologists in the later 19th century. It showed the ice sheets extending from northern Canada, southwards across New England and the Great Lakes region and westwards to the Rockies (a narrow corridor from an ice-free Alaska suggested a possible route for animals to have migrated from Asia into the rest of North America). The map also showed Greenland covered by another great ice sheet (as it still is), which helped to make the Ice Age—however dramatic—seem not wholly dissimilar to the present world.

Earth: in place of a singular mega-tsunami, a singular "catastrophic" drop in global temperature followed by its recovery. But the earlier field evidence that there had been more than one diluvial episode could readily be re-interpreted as evidence for more than one glacial episode. This was confirmed when, in the 1870s and in both Europe and North America, fossil soils and even the remains of forest trees were found sandwiched between separate deposits of glacial till or boulder clay. Intensive fieldwork, particularly by German and Austrian geologists, led to the reconstruction of repeated advances and retreats of the ice sheets: not only those that had extended southwards from Scandinavia but also those that had pushed northwards from the Alps (this belatedly vindicated Charpentier's reconstruction of the Rhône mega-glacier, setting it in a wider context). For example, Albrecht Penck, in *The Alps in the Ice Age* (*Die Alpen im Eiszeitalter*, 1901–9), synthesized a mass of late 19th-century fieldwork in this region. He described deposits formed during no fewer than four distinct glacial periods—which he named, in order, after the Austrian rivers Günz, Mindel, Riss, and Würm—separated by three "*interglacial*" periods, one of which, judging by its fossils, had been warmer in climate than Europe is today. This and similar research elsewhere in Europe and in North America demonstrated the sheer eventfulness of the Earth's history during the Pleistocene period. Its deposits might seem a mere top dressing overlying the thick Tertiary rock formations; but even the Pleistocene evidently represented an extremely lengthy span of time.

However, the sense of historical contingency that the unexpected Ice Age reinforced among geologists did not inhibit others from searching for its possible physical *cause*, especially when it became clear that there had been more than one glacial period within the total Ice Age. Significantly, this search started outside the bounds of geology itself. The self-educated Scotsman James Croll was one of the first to bring into the debate the research of astronomers on the Earth's measurable long-term variables, such as its orbital eccentricity and the precession of the equinoxes. In *Climate and Time* (1875) Croll argued that drastic glacial periods could have been triggered repeatedly by relatively small periodic changes in the Earth's motion in relation to the Sun. His theory of *cyclic* climatic change, with regularly repeated glacial periods, seemed at first quite plausible. But North American geologists concluded that his calculation of about 80,000 years since the last glacial period ended was incompatible with their own field evidence, which suggested that the last retreat of the ice sheets had been much more recent than this.

By the end of the 19th century Croll's theory was generally discredited. The cause of the Pleistocene Ice Age then remained as obscure as before, but its reality as a major event or sequence of events in the Earth's history was firmly established. Not for the first or last time, geologists recognized the crucial distinction between establishing that something had in fact happened in the past and finding a satisfactory explanation of *how* it had happened: in effect, between doing history and doing physics. But the sheer complexity of Pleistocene history brought it, perhaps paradoxically, more into line with the rest of the Earth's history. In effect, it tamed the Ice Age, turning it into the Pleistocene period, which in principle was as knowable and understandable as any other. (The time since the end of the Ice Age was distinguished as "*Holocene*" or "wholly recent"; Pleistocene and Holocene were together termed "*Quaternary*," since both were by definition post-Tertiary.)

## MEN AMONG THE MAMMOTHS

A Pleistocene Ice Age posed new questions for other puzzles about the relatively recent history of the Earth. How were these major climatic changes related to the apparent mass extinction of the mammoths and other large mammals that Cuvier had first reconstructed? Had they been killed off by the glacial conditions? Or had they been well adapted to the cold, as the mammoth's wooly coat suggested? And how was the Ice Age related to the origin and early history of human beings? Had the first humans been the contemporaries of the mammoths, or had they made their first appearance only after the Ice Age finally ended and the mammoths were gone? In the longer perspective of the Earth's total history, did the Ice Age or Pleistocene mark the boundary between the human world and the pre-human? If it did, it was from a human point of view surely the most decisive point in that entire story, and the most important period to understand.

Here a brief flashback to the earlier part of the century is needed, to give context to some sensational later developments. Claims for the contemporaneity of men and mammoths had been met with deep scepticism, notably and most influentially by Cuvier, who had good reason to doubt the authenticity of all the alleged finds of human fossils initially known to him. He remained sceptical, even when human bones and those of the extinct mammals were found close together, because there was often good evidence (as in the case of Buckland's "Red Lady" in Paviland cave) that they were not of the same age. He was convinced that the mass extinction of the large mammals had been caused not by any human activity but by a *natural* catastrophe of some kind; and this implied that it must have happened before human beings first came on the scene, or at least before they became a significant factor in the physical world. But in his later years Cuvier's justifiable scepticism hardened into unjustified dogmatism, in the face of increasingly reliable reports of genuine human fossils. For example, in several caves in the south of France the young naturalists Jules de Christol and Paul Tournal found a few human bones mixed with abundant animal bones preserved identically in the same deposits. Their reports polarized scientific opinion, and not only in France.

The best such case, however, was not reported until shortly after Cuvier's death in 1832. The physician Philippe-Charles Schmerling described caves in the Meuse valley near his home in Liège (in the then newly independent Belgium), and reported finding two human skulls—one of them lying close to a mammoth tooth—together with chipped flints and artifacts of bone. These were mixed with the bones of assorted extinct mammals, all buried in deposits deep beneath the floors of the caves and all preserved in just the same way. Schmerling was well aware of the doubts expressed by Cuvier and others about the earlier discoveries, and emphasized that he himself had excavated the crucial specimens with the greatest care: he insisted that there was no evidence that the human remains had been buried at some later time than the animal bones.

Yet even this powerful evidence failed to convince other geologists. Lyell, for example, visited Schmerling, saw his specimens and conceded that the case was "far more difficult to get over" than any previous one; yet get over it he did. He remained deeply resistant to anything that might be claimed (though not by Schmerling) as a new "man a witness of the Flood" in support of the biblical record. Yet Buckland himself, who was also among Schmerling's visitors, was still so conscious of his deceptive "Red Lady" that he too dismissed

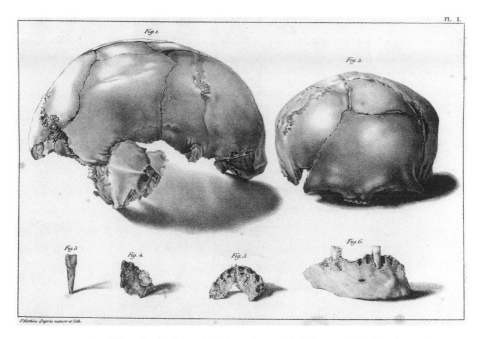

FIG. 8.2 Philippe-Charles Schmerling's illustrations (1833) of a human skull and jaw fragments, which he found associated with chipped flint implements and what he interpreted as animal bones fashioned by humans. All were found with the bones and teeth of mammoths and other extinct animals in a deposit beneath the floor of caves in the Meuse valley in Belgium. He claimed that this was clear evidence that humans coexisted with the extinct mammals during the diluvial period (re-interpreted by others, a decade later, as the Pleistocene Ice Age), but his contemporaries were generally sceptical.

the new case as likely to have been a similar human burial of a much later date than the fossil mammals. When other geologists remained sceptical without even visiting the site, Schmerling predicted bitterly that such "museum men" and "theory men" would one day be proved mistaken: as Tournal had pointed out, only the most careful *fieldwork* would ever persuade them otherwise, and only if they first gave up their dogmatic certainty that there could have been no overlap in time between human beings and the extinct mammals. Sadly, Schmerling died soon afterwards, long before his prescience was vindicated.

Christol and Tournal had in fact given the issue of human antiquity a crucially important new dimension. They suggested that their respective cave fossils dated from quite different times *within* the contentious diluvial period (as it was still being called at this point). Christol's caves (near Montpellier) had bones of many of the usual

extinct species; when Buckland visited him, he agreed that one cave
was a former hyenas' den just like the one at Kirkdale. In contrast,
the bones in Tournal's caves (near Narbonne) included those of several species known alive. Tournal therefore suggested that his cave
bones were *intermediate* in age. Human beings might have occupied
southern France without interruption from the period of the extinct
mammals through into historical times, while the mammal fauna was
changing gradually by the piecemeal extinction of some of the species,
perhaps as a result of human activities such as hunting or the clear-
ing of forests. This reconstruction left no place for any unique Deluge
event, and Tournal dropped his earlier use of the term "diluvial." He
argued that the human period at the tail end of the Earth's total history
comprised *two* periods of unequal length: the brief period of recorded
or literate human history, and a much longer preceding period of *prehistory* (he called it "*antehistoire*").

The idea of prehistory was not new: historians had used the word
ever since the late 18th century, but only to define what was in effect
unknowable because it preceded the earliest written records (those of
the earliest Egyptian and Chinese dynasties). What *could* be known
had been equated tacitly with the history of *literate* cultures. Tournal's
specific suggestion about a knowable but pre-literate period was not
widely noticed at the time—he was both young and a provincial—but
in the long run his idea was of the utmost importance. Just as Cuvier
had aspired to "burst the limits of time" by making the pre-human history of the Earth knowable, so human history could now be knowable,
back into the *pre-literate* times for which the only evidence was that of
human bones and artifacts. The well-established science of "*archaeology*" had in practice been focused mainly on the material remains of
ancient literate cultures (as for example in the excavations at Pompeii); but in the middle decades of the 19th century a new science of
"*prehistoric archaeology*" emerged, devoted to the study of pre-literate
human history.

That any such history was knowable and that it could perhaps be
reconstructed with some confidence were quite novel ideas. They were
ideas that were based, rather obviously, on what geologists had been
doing with increasing success in the preceding decades. Geology was
therefore repaying its earlier debt to the study of human history. Just
as the methods of human history had been fruitfully transposed into

the natural realm, so the methods of geology were now being applied to the study of the earliest human history; it was no coincidence that many of the leaders of the new prehistoric archaeology were originally geologists. Prehistory opened up a new conceptual space, as it were, between the pre-human history of the Earth, as newly reconstructed by geologists, and the literate human history traditionally described by historians. It had the potential to supply the link between the two and to unify them into a single historical narrative.

This was the context for a sensational breakthrough in the middle of the 19th century, which first established the historical reality of a lengthy Stone Age that had overlapped with the age of the extinct Pleistocene mammals. Schmerling's sad experience showed that the sceptics might never be persuaded that men had indeed lived among the mammoths, if the evidence was confined to caves. Cave deposits were always likely to have been disturbed, for example by burials such as that of Buckland's "Red Lady," because caves had been desirable dwellings for as long as human beings were around (some of them were still, in 19th-century Europe, the homes of "troglodytes"). Such evidence would only be decisive if caves were excavated with much greater care and precision than ever before. Alternatively, better evidence for the contemporaneity of men and mammoths might be sought well away from any caves, in the deposits that had formerly been classed as diluvium but had now been re-interpreted as Pleistocene river gravels. As it turned out, the breakthrough involved sites of both kinds.

In the valley of the Somme not far from the north French coast, several local antiquarians had found and described a variety of prehistoric stone artifacts, generally on the surface and therefore impossible to date. But in the 1840s the provincial civil servant Jacques Boucher de Perthes claimed to have found some such artifacts deep in the gravel pits near his home in Abbeville. The gravels had already yielded abundant fossil bones of some of the usual extinct mammals, so this was clearly a claim for contemporaneity. However, Boucher's *Celtic and Antediluvian Antiquities* (*Antiquités Celtiques et Antédiluviennes*, 1847) interpreted his finds in such a way that his claim was bound to be rejected by both geologists and archaeologists. He attributed the stone tools not to ordinary human beings, such as the "Celts" who had been living there when the Romans arrived, but to an antediluvian race of

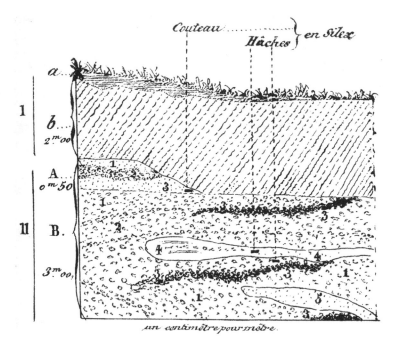

FIG. 8.3 Boucher de Perthes's section (1847) of one of the gravel pits near Abbeville, showing the exact positions of a flint knife (*Couteau*) and two flint axes (*Hâches*) found embedded in the same Pleistocene strata as the bones of extinct mammals, several meters below the surface. Although initially regarded by most geologists and archaeologists as suspect, such evidence for the "antiquity of Man" was later accepted as decisive.

pre-Adamites that had been totally annihilated, along with the large mammals, in the biblical Flood. Such theorizing was bound to seem to his readers a throwback to the age of Woodward and Scheuchzer, not a proper product of their own more enlightened times. It made it easy for them to dismiss Boucher as an ignorant provincial. And although he followed good geological practice by recording the exact position of the flint tools in the sequence of gravel deposits, many of his artifacts were suspect because they had been found not by him but by the workers, who might well have misled him or even deceived him deliberately. To cap it all, not a single human bone had been found among all the animal bones.

Unsurprisingly, Boucher's claims were either rejected or ignored, even by those geologists who were open to the possibility of the co-existence of early humans and the extinct mammals. But one well-qualified sceptic changed his mind as a result of making similar discoveries further up the Somme valley. The physician Marcel-Jérôme Rigollot, who had earlier been one of Cuvier's provincial informants on fossil bones, found flint tools *in situ* in bone-bearing gravel pits at St-Acheul, near his home in the city of Amiens. Sadly, he died before his published account had been fully discussed either by geologists or by archaeologists, and they all continued to consider the issue uncertain and unresolved. Boucher incorporated Rigollot's work in a second volume of his *Antiquities* (1857), which dropped many of his earlier and more fanciful interpretations and should have been more acceptable, yet it too failed to sway expert opinion in scientific centers such as Paris and London.

This logjam was breached soon afterwards by—surprisingly—evidence from a cave. Brixham Cave near Torquay on the south coast of England was discovered in 1858 and found to contain abundant fossil animal bones. English geologists quickly saw its potential: not to solve the problem of human antiquity, but rather to clarify the historical sequence of events by which the Pleistocene mammals had been replaced by modern species. The Geological Society raised funds that enabled the cave to be excavated with unprecedented care and in scrupulous detail, overseen by a distinguished committee that included archaeologists, geologists such as Lyell, and the anatomist Richard Owen (the "English Cuvier" who had first defined and named the dinosaurs). Under a crust of stalagmite precipitated subsequently (as in the caves at Kirkdale and in Bavaria), masses of animal bones were extracted, and precise records were kept of their places in the successive layers of cave deposit. Among them, as an unanticipated bonus, were a few indisputable chipped flint tools, clearly *in situ* beneath unbroken stalagmite. It seemed indisputable that their makers must have been alive at the same time as extinct species of hyena and rhinoceros.

However, even Brixham Cave was still not quite enough to allay geologists' long-standing worries about evidence derived from caves. And so, in the following months, several of those involved with Brixham, among them Lyell, crossed to France to visit Boucher and see for themselves his and Rigollot's gravel pits in the Somme valley. On the spot they were duly convinced that the French finds were genuine after all: seeing a newly discovered flint tool still embedded in the face of a gravel pit, deep below the surface, left little room for doubt. In

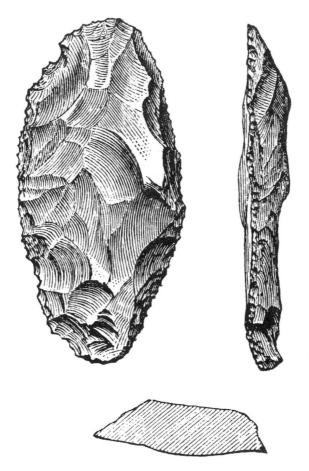

FIG. 8.4 One of the prehistoric chipped flint tools (two views, and a cross-section) found in 1858 among the bones of extinct mammals beneath the floor of Brixham Cave in southern England, as illustrated by the London geologist (and prosperous wine merchant) Joseph Prestwich in his published report on the excavation (1873). Such finds, recorded with unprecedented care and precision, helped to eliminate remaining doubts about the contemporaneity of human beings and the Pleistocene animal fauna. They proved that human history extended well back into what had previously been assumed to be *pre-human* times, even if the human species was still a *relative* newcomer in the total history of the Earth.

1859, in a coordinated campaign to change scientific opinion in Britain, they reported their conclusions at several scientific meetings, addressing geologists, archaeologists, and other "men of science." Lyell drew a *historical* conclusion: there had been "a vast lapse of ages, separating the era in which the fossil implements were framed and that of the invasion of Gaul [France] by the Romans." Human beings with a highly skilled technology—still being practiced in the 19th century, to make flints for flintlock firearms—had already been around in Europe during the Pleistocene Ice Age, or at least in its milder interglacial periods.

There was still some resistance to this conclusion in Paris, where Élie de Beaumont—a powerful figure in the Académie des Sciences—maintained Cuvier's sceptical stance. But other Frenchmen increasingly supported the rehabilitation of Boucher and all that it implied. However, it remained unclear what kind of human beings had fashioned the flint artifacts. This issue was further confused when at last the first human-like fossil was found in the Somme gravels. For the jawbone found in 1863 at Moulin-Quignon (near Abbeville) proved highly controversial. Most of the relevant French naturalists claimed it was authentic; but most of the English suspected it had been planted, probably by one of the workmen, some of whom were certainly engaged by this time in a profitable tourist trade in "repro" flint tools. The matter was debated in a scientific "trial of the jaw," with the leading experts from both nations presenting their arguments, first in Paris and then on the spot in Normandy. Although the case was formally decided in favor of its authenticity, doubts persisted and eventually the Moulin-Quignon jaw was rejected as fraudulent. So the character of the tool-making beings was as uncertain as ever.

The geologist Édouard Lartet, who had first proposed this international debate, argued that what was needed was a relative *chronology* for prehistory, to match what the geologists had already constructed for the deeper reaches of the Earth's history. Building on Tournal's earlier suggestions, he outlined a tentative sequence of four ages based on the successive sets of mammals that had coexisted (at least in Western Europe) with early human beings. These were the ages of cave bears, of elephants and rhinoceros, of reindeer, and lastly of aurochs (an extinct species of wild cattle). This was a sequence that had apparently seen the *piecemeal* extinctions of the earlier mammals, there being a strong suspicion that the early human beings might have been responsible.

An analogous sequence had already been suggested to account for the artifacts left by the earliest human cultures. In 1837 the Danish antiquarian Christian Thomsen had argued that they formed a sequence reflecting increasing levels of technical sophistication, from a Stone Age through a Bronze Age to an Iron Age. This *"three-age system"* was primarily an organizing principle for the classification and display of the varied artifacts in the museum for which Thomsen was responsible; it was based on a plausible assumption about human technical progress. But even the Stone Age, which was characterized by smoothly polished stone tools or weapons, was clearly more recent than the age in which the chipped flint implements had been made. John Lubbock, a young "man of science" (and banker) in London, therefore suggested in *Prehistoric Times* (1865) that Thomsen's earliest period should be renamed the *Neolithic* ("New Stone") Age, while the even earlier period of chipped stone tools should be called the *Palaeolithic* ("Old Stone") Age. Brixham Cave and the Somme gravels then became Palaeolithic sites, unquestionably far older than, say, prehistoric monuments such as Stonehenge. And in 1872 the Old Stone Age was itself divided by the French archaeologist Gabriel de Mortillet into a sequence of phases characterized by progressively more skillful chipping techniques (one of the earliest, the Acheulian, was named after Rigollot's locality of St-Acheul). It then seemed worthwhile to try to correlate these with Lartet's periods based on the changing sets of mammals among which these early human beings had lived.

At its earliest end, however, this tentative history of human activity was not straightforward. There was a long and heated argument over what were called "*eoliths*" ("dawn-stones"), flints with some signs of chipping but no unambiguous overall design, which were found in deposits dated by their fossils as Pliocene in age. If they were genuinely human artifacts, they would push human antiquity back even before the Pleistocene period. But rather similar flints with random chipping were also found on modern beaches and in modern rivers, and were certainly natural in origin. In the end most geologists concluded that eoliths were not the result of human workmanship, so traces of early human life were confined to the Pleistocene. But this was sensational enough.

During the rest of the 19th century the history of the Pleistocene period was reconstructed, at least for Western Europe, with increasing confidence. This integrated a story of alternating glacial and interglacial phases in the climate with one of changing faunas and of a sequence of slowly developing but apparently pre-literate human cultures. Lyell's *Antiquity of Man* (1863) synthesized what, even then, was well on the way to becoming a broad international consensus among geologists and archaeologists. He outlined a historical sequence stretching from the Pliocene period in the late Tertiary era—a pre-human world, if the contentious eoliths were discounted—through the Pleistocene period with its extinct mammals and early human beings, towards the post-glacial world and the times of recorded human history. By the end of the century, Lyell's putative "vast lapse of ages" had been filled with a sequence of human cultures that, at least in outline, linked the contemporaries of the extinct mammoths all the way to the Iron Age people whom the Romans had encountered when they colonized northern Europe and brought literate history to that region.

## THE QUESTION OF EVOLUTION

The establishment of human antiquity was dubbed by Murchison a "great and sudden revolution," which anchored the human species into the tail end of the Earth's much deeper pre-human history. It had its decisive moment in 1859, when, as already mentioned, leading "men of science" agreed that in the Somme gravels there were genuine flint tools alongside the bones of extinct mammals. But the year 1859 is now more famous as the year in which Darwin's book *On the Origin of Species* was first published. It might seem in retrospect that the long-term history of life cried out for explanation in terms of the gradual *evolution* of all the plants and animals—and finally the human beings—in the fossil record. In fact, however, an evolutionary explanation of this history was not widely accepted, either by "men of science" or by the educated public, until after Darwin's *Origin* began to make it look more plausible. From this it might seem easy to conclude that only the reactionary influence commonly attributed to "Religion" or "The Church" could have prevented an evolutionary interpretation from being put forward and accepted much earlier. But this would be a serious misreading of the relevant debates.

Backtracking briefly to the 18th century, at that time the descriptive sciences of "natural history" had barely any *historical* dimension at all, except in "Theories of the Earth" and similarly speculative works. In the everyday work of natural history, mineralogists classified minerals into "*species*" (kinds) in just the same way as botanists classified plants and zoologists classified animals. It made little sense in any of these sciences to ask about the *origins* of their species. The species called "daisy" and "lion" were taken to be permanent features of the diversity of the world, no less than the species called "quartz" and "salt." The question of their origins might be an important matter of metaphysics or, for some religious believers, a matter of God's actions at the very beginning of time (during the "days" of Creation). But the question seemed to be both inappropriate and unanswerable within the sciences themselves. Only with the development of the idea of a *history* of the Earth and its life (as sketched in earlier chapters of this book)—and most of all the idea of the reality of extinction, which first made past worlds distinctively different from the present—did it begin to make any sense to ask questions about the origins, *within* this history, of the diverse forms of life, living and extinct. The origin of species then became a problem within the sciences, and a major one at that. The leading "man of science" John Herschel later called it "the mystery of mysteries," but that meant that it was an outstandingly important puzzle waiting to be solved, not that there was any reason for it to remain a mystery forever.

Only around the start of the 19th century, then, did it make any sense to try, in a more than merely speculative manner, to construct theories of what is now called evolution. However, as mentioned in an earlier chapter, Lamarck's basic idea had been that all forms of life are perpetually changing or "*transmuting*," although extremely slowly, so that "species" are ultimately unreal and arbitrary, and their origin from preceding species is just a matter of the passage of time. Yet in practice, in his fine work on the fossil shells from the Tertiary formations around Paris, Lamarck had treated molluscan species as real natural units, just as his younger colleague Cuvier was treating those of living and fossil mammals. There had been an almost complete disjunction between Lamarck's evolutionary theorizing and his practical work in describing and naming species. This highlights one of the strongest reasons for the reluctance of most other naturalists early in the 19th century to concede that species might have evolved slowly from others in the course of time. They found in practice that, with only a few troublesome exceptions, species were clearly distinct from one another. For example, as Cuvier had famously demonstrated just before the start of the century, the Indian elephant was distinct from the African species, and both were distinct from the extinct mammoth. And Lyell's ingenious timescale for the Tertiary era, plotting the changing percentages of extinct and extant species of fossil molluscs, depended crucially on the reality of those species as distinct natural units that could be counted. In sum, the spectacular development of geology had failed to turn up any good fossil evidence for the gradual transmutation or *evolution* of any one species into another in the course of geological time, as Lamarck's theory proposed. Later theories had to explain this, or explain it away.

An alternative kind of theory, which avoided this problem, conceded the reality of species as natural kinds that remained unchanged for as long as they existed at all, but suggested that a new species might arise from an earlier one by some relatively sudden change (somewhat like the modern evolutionary concept of "punctuated equilibrium"). Brocchi's idea of the piecemeal "births" of species in the course of time had implied this possibility. And Étienne Geoffroy Saint-Hilaire—a zoologist who was Lamarck's ally and Cuvier's critic at the great Muséum in Paris—developed such a theory more explicitly in the 1820s and 1830s, basing it on the occasional and apparently random appearance of "*monstrosities*" in chicks hatched at a chick-rearing factory and, far more distressingly, in human births at the great Parisian hospitals. Geoffroy suggested that new species might appear in the natural world in rather the same way, as "hopeful monsters" (the term used by its critics to characterize a somewhat similar 20th-century theory). He also claimed that living Indian crocodiles (gavials) were derived "by way of uninterrupted generation" from the distinctly different fossil crocodiles found in Jurassic strata, but his suggested line of descent made nonsense of the fossil evidence and was easy to dismiss. Anyway, this kind of theory was unacceptable or even repugnant to most naturalists because it failed to account for the apparently designful adaptations of organisms to their specific ways of life (whether or not this was attributed to divine providence): it made adaptation a matter of unlikely chance. Nonetheless, similar conjectures about the possible origins of new species, by some process involving natural "*saltations*" or "leaps," continued to be widely debated, particularly in mainland Europe. And many other possible models for organic change were discussed or at least hinted at, by naturalists who for whatever reason found Lamarck's kind of theorizing unsatisfactory. It is only in hindsight—and partly as a result of Darwin's eloquent rhetoric in his *Origin*—that his particular kind of evolutionary theory, with all the

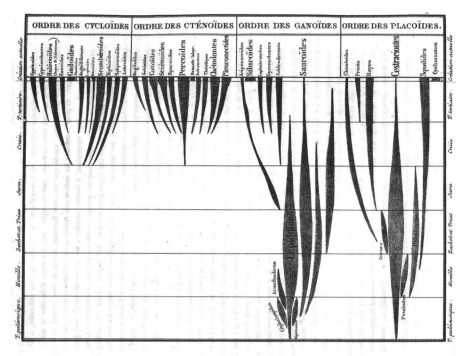

**FIG. 8.5** Louis Agassiz's "Genealogy of the Class of Fishes," published in 1843 in his great *Researches on Fossil Fishes* (*Recherches sur les Poissons Fossiles*, 1833–43). Time flows upwards, as in the pile of formations (named in the left and right margins) on which the diagram was based. The breadth of each "spindle" represents his impression of the relative abundance and diversity, through time, of each group of fishes. Two of his four major "orders" (*Ganoids* and *Placoids*) were abundant in earlier periods; the other two (*Cycloids* and *Ctenoids*) first appeared in the Cretaceous period (*Craie*) and became diverse only in the Tertiary era (*Terrain tertiaire*) and the present world (*Création actuelle*, at top). Despite its striking similarity to some modern evolutionary diagrams—and his use of the word "genealogy"—Agassiz was adamantly opposed to Lamarck's kind of evolutionary theory and, later, to Darwin's.

emphasis on extremely slow and gradual change, may seem to have been the only one available for discussion in the 19th century. Darwin claimed that the only alternative to his theory was to invoke the sudden appearance of new species by miraculous divine intervention. It was not. Many naturalists were thinking about other possible ways in which new species might arise by natural means, though none was as well developed as Darwin's theory.

Darwin first made his name in the scientific world *as a geologist*. He was profoundly influenced by Lyell, whose *Principles* he had with him during his subsequently famous voyage on the *Beagle* (he was its unofficial naturalist and the captain's social companion during its official hydrographic survey of the South American coastline, but he spent as much time as possible on land). Before the voyage he had briefly been taught geological fieldwork by Sedgwick; on his return he became an active member of the Geological Society, described himself to his fiancée as "I, the geologist," and spent the following years writing and publishing geological papers and books arising from what he had seen in the field in the course of the voyage. At the same time, however, he was privately developing his evolutionary theory, well aware that it would have to be much better grounded in detailed evidence than those previously suggested by others, if it was to have any chance of being accepted by "men of science," let alone by the wider public. He devoted eight years to an exhaustive and exhausting study of living and fossil barnacles, so that no future critic could say that he lacked first-hand experience of the species problem. Emulating Lyell's *Principles*, he planned to write a massive work entitled *Natural Selection*, which was his name for what he proposed as the main (but not only) cause of evolutionary change. *On the Origin of Species*, the title of the still quite massive "abstract" that he eventually published in 1859, likewise expressed his intention to focus his theory on the strictly limited *causal* problem of how any new species could evolve from an earlier one, rather than trying to reconstruct the complex course that evolutionary change might have taken during the Earth's lengthy *history* (his title referred to the origin of species, in the plural, not, as it is often misquoted, "the origin of *the* species" in the singular).

Darwin had to explain away the awkward fact that there was no hard evidence in the fossil record that evolutionary changes had been due to the extremely slow and gradual process that he proposed. But he was fully convinced by Lyell's insistence on the vastness of the deep time that had been available, combined with Lyell's more controversial argument about the extremely imperfect character of the fossil record. Darwin also adopted his mentor's "uniformitarian" or steady-state model, at least for the history of the Earth if not for that of its life. Like Lyell, he argued that the Earth's crust is composed of huge crustal plates forever oscillating slowly up and down (not shifting sideways like those of modern plate tectonics). He used this theory to explain, for example, the varied forms of coral reefs: the apparently distinct kinds ("species") of fringing reefs, offshore reefs, and atolls were just successive phases in a continuous process by which the coral organisms maintained themselves near sea level while a crustal plate was sinking slowly beneath them. This offered a useful analogy with his idea of an equally slow and continuous process of organic evolution; it also suggested how an ever-changing geography could provide all the ever-changing environments in which new species could diverge slowly from their ancestral forms.

During the rest of the 19th century, biologists argued endlessly about the *causes* of evolutionary change. Darwin's idea of natural selection was somewhat eclipsed and came to be regarded as probably just one cause among several. But most palaeontologists simply got on with interpreting the fossil record in terms of the evolutionary *history* of life, accepting that they could say little about the causes of evolutionary change. And while biologists enthusiastically constructed "family trees" to show the supposed ancestry of living organisms— often simply converting systems of classification into an evolutionary format—a growing number of palaeontologists tried, more realistically, to link up the fragmentary fossil evidence into reconstructions that might at least approximate the way that the organisms had *in fact* evolved in the course of the Earth's history.

This contrast explains how palaeontologists were able to accept quite readily the historical reality of evolutionary change, and incorporate it into their research, without having to become involved in the arguments about how the changes had been caused. They became *"evolutionists"* without having to choose (though some of them did) between joining the ranks of Darwinians or neo-Lamarckians or saltationists or any other evolutionary sect. It made much better sense of their fossils to infer that they were indeed related to one another by evolutionary connections—by "descent with modification," as Darwin put it—even if the fossil record was too fragmentary to show whether the changes had taken place gradually or suddenly, let alone to show how the changes had been caused.

Much more important, for swaying scientific opinion in favor of evolutionary change *however it had been caused*, were the occasional discoveries of new fossils that were interpreted as previously "missing links" between major groups of animals or plants. Among these

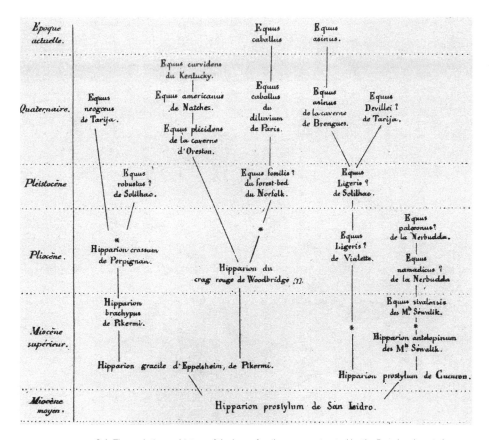

**FIG. 8.6** The evolutionary history of the horse family, as reconstructed by the French palaeontologist Albert Gaudry in 1866. Time flows upwards, as in the pile of Tertiary (and recently named *Quaternary*) formations in which the fossils were found. The two horse species—the horse proper and the ass—that are still alive at the present day (*Époque actuelle*, at top) are shown as survivors of a much more diverse "bush" of fossil species, some extinct and some that were ancestors of later species (asterisks denote the connections that Gaudry thought uncertain). This reconstruction was based on real fossil evidence: in each case the locality of the fossil species was noted (e.g., Paris, Pikermi in Greece, the Siwalik hills in India, Norfolk in England). In contrast, most evolutionary reconstructions in the 19th century, although rather similar in appearance as "genealogies," were based mainly or wholly on the supposed relationships of *living* organisms, and their extension into the deep past was largely hypothetical.

new finds, not only intermediate in their anatomy but also of about the right geological age, was the Jurassic *Archaeopteryx*, a plausible link between reptiles and birds. Although controversial, it was eventually taken to be strong evidence for the reality of evolutionary change between what are now quite distinct major groups (in modern terms, it illustrated a case of "*macro-evolution*"). In contrast, fossils failed

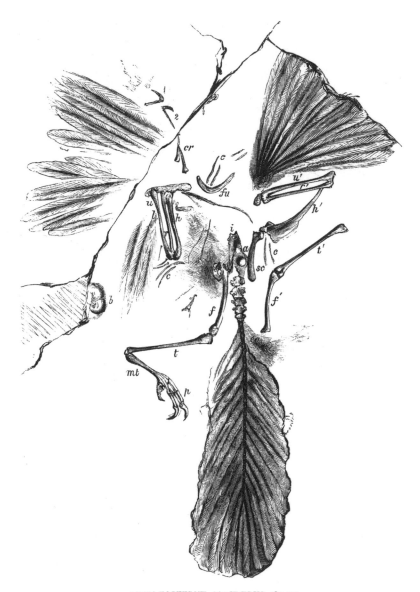

ARCHÆOPTERYX MACRURUS (Owen).

In the National Collection, British Museum.

**FIG. 8.7** The fossil *Archaeopteryx* ("ancient wing"), found in 1861—only two years after Darwin's *Origin of Species* was published—in the Jurassic limestone of Solnhofen in Bavaria. It was interpreted as a reptile-like bird, somewhat intermediate in its anatomy, and it was of about the right geological age to be a plausible link in the *evolutionary* origin of birds from a reptilian ancestor; a small dinosaur found at almost the same time in the same deposit appeared to be a somewhat bird-like reptile, which strengthened the case still further.

almost completely to provide clear evidence of the kind of small-scale change on which Darwin's *Origin* was focused, namely between one species and a very similar but later one ("*micro-evolution*"). Two such cases were notable for being almost the only ones of their kind. One, involving a sequence of Tertiary freshwater molluscs, was correctly described in 1875 by the Austrian palaeontologist Melchior Neumayr as "a contribution to the theory of descent." In the other, English palaeontologists traced "an unbroken continuity in the evolution of *Micraster*," a fossil sea-urchin found at many levels in the exceptionally uniform strata of the Chalk formation. But such cases were so rare that they did little to stem the growing doubts in the later 19th century about Darwin's specific brand of evolutionary theory.

## HUMAN EVOLUTION

Darwin's *Origin*, in successive editions up to his death in 1882, continued to be focused on the crucial but limited problem of how *any* new species might have been formed from a different ancestral one. Right from the start, however, he and other "men of science," and the educated public, were acutely aware that this had an obvious bearing on the question of the origin of what they all regarded as the most important species of all, and on the far more profound question of what it means to be human. As Darwin put it prophetically—at the only point in the *Origin* where he mentioned it at all—one of the results of his theory would be that "light will be thrown on the origin of man and his history." The question of *human* evolution was never far away.

Back in the 1820s, the young Lyell's *volte-face*—from interpreting the fossil record in terms of a broadly progressive sequence, to denying any such progressive element in favor of a steady-state or cyclic system (the dinosaurs might *return*)—was probably triggered by his first encounter with the evolutionary ideas of Lamarck and Geoffroy. As he realized at once, a "progressive" fossil record in which the first fish had been followed much later by the first reptiles and later still by the first mammals could all too easily be turned into an evolutionary sequence and then be extended to the first humans. Conceding the natural origin of *Homo sapiens* by evolution from some kind of ape—or, as he put it, "going the whole Orang[-Utan]"—threatened what he and many others regarded as the irreducible "dignity of Man." This was a crucial issue on which Lyell the deist saw eye to eye with, for example, Sedgwick the Christian theist. For neither man was it a question of a threat to biblical literalism. Far more importantly, human evolution seemed to threaten what it meant to be a morally responsible human being rather than an amoral animal. This explains why, decades later, the elderly Sedgwick so vehemently rejected what his former student had published as the *Origin*, while yet signing off, "[in] spite of our disagreement in some points of the deepest moral interest, your true-hearted old friend." Lyell, in his *Antiquity of Man*, did abandon some of his earlier ideas, but his lukewarm and qualified acceptance of evolution disappointed Darwin. Many savants in the later 19th century, including Lyell, came to accept the physical evolution of their own species, but balked at Darwin's purely physical explanation for the origin of the moral sense, of conscience and consciousness, and of all that seemed to make human beings fully human.

Meanwhile the human species had become ever more closely integrated into the fossil record. In the early 19th century the primates (the group of mammals to which, anatomically, human beings belong) had been strangely missing from even the largest collections of fossils. In 1837, however, not long after Schmerling had published his controversial finds of human fossils from Belgium, fossil bones of non-human primates were reported—by coincidence, almost simultaneously—from Tertiary deposits in southern France, the Himalayan foothills, and Brazil. The conjunction became the occasion for a striking reconstruction of a conjectural human ancestor. Although not a piece of serious science, this gratuitously bestial image showed all too clearly that Lyell and Sedgwick were later quite justified in worrying about where evolutionary theorizing might lead (and the strident reductionist claims of some modern ultra-Darwinists, that humans are in effect "nothing but" naked apes, suggests that little has changed in this respect since the 19th century).

More direct fossil evidence for the evolution of human beings from some kind of less-than-human primate was slow in coming. A skull that was clearly human-like but not *Homo sapiens* was found in 1856 in a cave in the Neander valley near Dusseldorf in Germany; but as usual with cave finds the geological age of "Neanderthal Man" was highly uncertain, and anyway some anatomists thought it was no more than a

FIG. 8.8  "Fossil Man": a human ancestor as imagined in 1838 by Pierre Boitard, a French author of popular science, for a magazine article on the history of life. This was probably the earliest example of what became (and has remained) a popular kind of scene from deep time. The negroid or even simian features show why the "dignity of Man" could be regarded as threatened—not least in the fiercely contested 19th-century politics of race—when the then new evidence of fossil primates was combined with the still ambiguous evidence of human fossils. This highly conjectural image was recycled half a century later, in 1887, by the German evolutionist Ernst Haeckel, to illustrate his conception of his hypothetical "missing link" *Pithecanthropus* ("ape-man").

pathological aberration. After Darwin's *Origin* was published, his supporter the zoologist Thomas Huxley reviewed *Man's Place in Nature* (1863) and rehabilitated Schmerling's Belgian skull as a genuine human fossil. But he judged it "a fair average human skull, which might have belonged to a philosopher [such as himself!], or might have contained the thoughtless brains of a savage," so that it could throw no light on the origin of the human species. When Darwin himself explicitly extended his evolutionary argument in *The Descent of Man* (1871), he could not point to any clear *fossil* evidence for human evolution, though he had persuasive evidence of many other kinds.

Not until 1882 were Neanderthal skulls of unambiguously high antiquity found, but by then there were good anatomical reasons to doubt if that species was our direct ancestor. Only in 1891 was a better fossil candidate found for *Pithecanthropus* ("Ape-Man"), the hypothetical "missing link" that evolutionists had already confidently predicted. This was the "Java Man" or *Pithecanthropus erectus* (now *Homo erectus*) found by the Dutch biologist Eugène Dubois in the Dutch East Indies (now Indonesia). From that point onwards, whatever the arguments—which were many and heated—about the course of human evolution, *Homo sapiens* and its putative ancestors were all firmly anchored into the tail end of the total history of life, or, more commonly, were treated as its culmination. The next chapter remains in the later 19th century, but returns to the wider issues of this total history of life, seen as a part of the long-term history of the Earth itself.

# 9

# Eventful Deep History

## "GEOLOGY AND GENESIS" MARGINALIZED

In the second half of the 19th century there were still certain parts of the religious public that objected strongly to the geologists' idea of an extremely ancient Earth—with human beings appearing on the scene as it were at the last moment—on the grounds that it was incompatible with the plain and literal meaning of the opening chapters of Genesis. But such views were being edged out to the margins of intellectual life. They continued to be confined almost entirely to the English-speaking world and, even there, mainly to the less well educated. Others noticed that religious people, some of them ordained clergy, were well represented among the leading geologists. This reinforced their sense that geology was clearly compatible with religious practice, while weakening the contrary claims of some secularists that this science, even more than others, was undermining religious belief altogether.

Since the Bible was central to Christian worship and to the worldview it sustained, particularly in its Protestant forms, much religious discussion continued to focus on biblical interpretation. But any notion of a single unambiguously "literal" reading of the Genesis narratives—or any other biblical text—

had already been weakened from two directions. Back in the 18th century, in the movement of the Enlightenment, the study of the Bible had been transformed by the adoption of *historical* methods of interpreting ancient texts of any kind: they needed to be understood in terms of the cultures in which and for which they were originally composed, at particular periods in the past. And in the case of biblical texts, the Romantic movement in the early 19th century had often emphasized their *literary* character and their pervasive use of metaphor, analogy, symbolism, and poetry, while recognizing that this could enhance rather than constrain their capacity to express profound ideas. Influenced in these ways, biblical criticism had proved to be a two-edged sword. It could indeed be used to undermine or even allegedly demolish the value of the Bible, for example in the service of radical political objectives, but alternatively it could stimulate theological understanding in the service of a deepened religious practice. In Britain, such issues came belatedly to a head with the publication of *Essays and Reviews* (1860), a book that in its sales and its immediate impact far exceeded Darwin's *Origin* the previous year. It gave the educated public in Britain an overdue taste of theological currents that had long been swirling around intellectual centers in the rest of Europe. Predictably its authors attracted vehement criticism from some religious traditionalists, but many other readers found the book a liberating experience. By the end of the century, among educated people throughout Europe—the religious as well as the sceptics—the kind of naïve literalism epitomized by assertions that "the Bible says . . ." had come to be regarded as clearly untenable.

On the specific issues related to the science of geology, this deepened understanding of the character of biblical texts affected interpretations of all the relevant narratives in Genesis. For Noah's Flood or Deluge, there had been a succession of attempts (as outlined in earlier chapters) to identify features in the natural world that might be traces of such a drastic historical event. At first it had been held responsible for all the Secondary rock strata and all their fossils, but later just for the Superficial or diluvial deposits. When these in turn were re-interpreted as traces of the Pleistocene glaciations, the Genesis account was further restricted—if it had any historical foundation at all—to a relatively local or regional event, some time around the end of the Ice Age or even more recently. These changes in interpretation were welcomed by sceptics and atheists as debunking the biblical text altogether, while religious conservatives denounced them for abandoning the revealed truth of the story. But what in fact they illustrated was the *historicizing* of Genesis: the universal extent of the inundation, for example, was now taken to refer to the world as it was known to, and understood by, those for whom the narrative was first composed. The imputed religious *meaning* of the story, if believers wanted to retain it, as many did, was little altered.

This historicized interpretation of the Flood story in Genesis was reinforced in the later 19th century, as an unexpected result of the successful deciphering of the ancient cuneiform inscriptions that were being found by archaeologists in Mesopotamia. One leading expert on cuneiform, George Smith, was working for the British Museum on the hundreds of clay tablets that had been excavated at sites such as Nineveh (near Mosul in modern Iraq). In 1872 he reported that he had found a tablet recording a Flood story strikingly similar to the biblical narrative, with Izdubar paralleling Noah. But far from treating this sensational discovery as evidence that the story in Genesis must be derived from an earlier source, Smith suggested that the two accounts were parallel records of the same event; he explained their differences in terms of the contrasting physical environments of Mesopotamia and Palestine, the regions in which the stories had first been composed. So while some commentators argued that his find finally demolished any claim that the biblical text was uniquely inspired, others noted that what the biblical version had done was to give the story a distinctive *religious* interpretation in line with other parts of the Jewish scriptures: for example, it ended with a divine assurance that such a catastrophe would never again be inflicted on humankind. In any case the discovery led scholars, and the wider educated public, to conclude that the biblical Flood had not been literally global but probably confined to Mesopotamia. Yet it could still have been a genuine historical event, a disastrous regional inundation for which physical evidence might still turn up somewhere in the Middle East. Smith later deciphered other tablets recording a Mesopotamian account of the Creation and Fall, much fuller than the biblical version. This suggested that the latter represented a selective reworking of earlier material; again, this had accentuated its *religious* character in line with Jewish ideas, for example by describing Creation as the unaided

FIG. 9.1 The clay tablet recording the ancient Mesopotamian story of the Flood, as reassembled by George Smith from fragments found in the excavations at Nineveh. This engraving was published in his *Chaldean Account of Genesis* (1872); it was intended just to give a general impression of the tablet, and at this small scale the cuneiform characters are not reproduced with any accuracy.

work of a uniquely transcendent Creator who had pronounced each new feature of the world intrinsically "good."

Meanwhile the biblical account of the six "days" of Creation already had a long history of being interpreted in terms of its possible traces in the rocks (again, as outlined in earlier chapters). In the 19th century the rapidly developing science of geology was providing new and cumulative evidence for a directional and broadly "progressive" history of the Earth and its life (seriously contested only by Lyell). A wide variety of commentators, including some competent geologists, eagerly matched or "harmonized" this sequence of geological periods with the "days" of the Genesis narrative. The two accounts, biblical and geological, were quite readily "reconciled," provided that it was conceded, as biblical scholarship indicated, that the Hebrew word translated as "day" might in this context denote a moment or period of divine significance rather than a span of twenty-four hours. Some other commentators, unwilling to make that concession yet not wishing to reject the new scientific discoveries altogether, suggested—somewhat desperately—that the entire geological history could be inserted into a hypothetical gap in the biblical narrative immediately after the initial act of creation and before even the first of the six "days" (all of which could then be retained as ordinary days). In any case, on this issue of the primal Creation, as on that of the Flood, the incompatibilities that some of the earlier "scriptural" writers had claimed to find between geology and Genesis ebbed away in the later 19th century; or at least they retreated or were relegated to the social, intellectual, and religious margins of society. When, for example, Philip Gosse, a competent English naturalist but also a member of the extremely conservative sect of the Plymouth Brethren, published a book entitled *Omphalos* (1857), its ingenious—but unfalsifiable—argument against evolution and in favor of a "young Earth" was dismissed by religious and non-religious people alike (the title referred to Adam's presumed possession of a navel or *omphalos*, despite having supposedly been created directly from "the dust of the ground" rather than having a normal birth).

In effect, geology and Genesis became dissociated in the course of the 19th century, and, in general, amicably so. Certainly the many geologists who were practicing Christians did not find their faith undermined *by their science*, though some of them had other reasons, often ethical ones, for abandoning the kind of Christianity that was dominant in their culture (Darwin was one of these). But just as specific social and political conditions had generated a brief heyday for "scriptural" writers in Britain early in the century, so towards its end, under equally specific circumstances, there was a surprising late revival of biblical literalism in the United States. In 1881, for example, Archibald Hodge and Benjamin Warfield, theologians at the Presbyterian seminary at Princeton, published an influential article on the divine "inspiration" of the Bible, in which they asserted that "all the affirmations of Scripture . . . are without any error." Since matters of "physical or historical fact" were included, this implied that in the case of any discrepancy with scientific ideas the Bible would be given the last word. But this verbal "*inerrancy*" was said to reside not in the Bible as it was read in church or at home, but in its "original autographs" or manuscript texts. These were of course no longer available to anyone, and could only be reconstructed from the extant texts by deploying the usual scholarly tools of biblical criticism. So this startling *and novel* claim of absolute verbal inerrancy was both powerful and yet unfalsifiable. Although it was strongly criticized by other theologians, even as "theological rubbish," it was embraced enthusiastically by major factions in American Protestantism, as a valuable weapon with which they could confront what they perceived as the secularizing forces of modernity. This was not the first or the last time that American culture exhibited its "exceptionalism" or—in both senses—its peculiarity.

However, what remained far more widespread in the 19th century, throughout the Western world, was a confident sense that the geologists' new scientific knowledge was confirming and even reinforcing the belief in the overarching and providential designfulness of nature. This kind of natural theology had earlier characterized, for example, not only Buckland's explicit Christian theism but also Lyell's implicit deism. And Buckland had convincingly extended it into the deepest history of life, by showing that even the earliest known organisms were well designed to fit them for specific modes of life. But later in the century, in the wake of Darwin's *Origin*, this widespread sense of providential design in nature was rightly seen to be threatened by evolutionary ideas. For Darwin proposed a radical alternative explanation of the apparent designfulness of organisms—as a purely natural product of natural selection—which dramatically undercut the plausibility of this part of the traditional "argument from design." And natural selection itself was widely thought to make the entire evolutionary process a product of blind chance, which was particularly objectionable because Darwin extended it to human beings. However, this erosive effect of Darwin's theory was uneven in its impact. It was felt most strongly in those cultures—notably his own in England—in which the intellectual defense of Christian faith had become most heavily dependent on natural theology, and especially on the designfulness of the natural world, rather than on the religious significance of the specific historical events that had always been at the heart of the mainstream Christian worldview. Anyway, these issues had little impact on the everyday work of geologists, whether religious or not.

## THE EARTH'S HISTORY IN PERSPECTIVE

For much of the 19th century that everyday work lay mostly among the formations and fossils that had formerly been classed as Secondary, together with the more recently unraveled Transition ones that extended the Earth's history still further back into the deep past. What were more problematic and difficult to understand were not only the youngest or "diluvial" deposits—until they were re-interpreted as traces of an Ice Age—but also, less surprisingly, the Primaries or oldest rocks of all, which lacked fossils of any kind. It was in between, in the Secondary and Transition formations, that geologists made their most striking progress. The rapid improvement in their detailed knowledge of the sequence of formations and their fossils confirmed and consolidated their picture of the history of the Earth and its life as having been broadly directional and even progressive, while Lyell's contrary picture of "absolute uniformity" became less and less plausible. Yet the value of both pictures can be seen in what would otherwise be paradoxical: all this research revealed the deep past as having been both unfamiliar and familiar, both strange and ordinary.

Dinosaurs and trilobites were certainly strange, and often evoked feelings of wonder, which were exploited to the full in promoting the science for a wider public. When the world's first great international exhibition was moved from central London to a more permanent site in the suburbs, its grounds were adorned with a new outdoor exhibit that was both entertaining and instructive. Under the scientific direction of Owen, who had first defined and named the dinosaurs, they and other spectacular extinct creatures were represented by life-sized reconstructions, arranged to show their correct sequence in the total history of life. Later in the century, exploration of the American West, particularly in the wake of the new railways that were opening up the vast continent, led to the discovery of the fossil bones of spectacular new dinosaurs and extinct mammals. Reconstructions of them, following the methods pioneered long before by Cuvier, soon became (and have of course remained) some of the most prominent exhibits in natural history museums in North America and around the world. The strange otherness of the deep past was put on permanent display.

In these same decades, however, geologists also found cumulative evidence for the relative ordinariness of the deep past. Even the strangest extinct animals and plants had lived in environments that were increasingly recognized as quite familiar. Earlier naturalists had, for example, described extinct volcanoes and their lava flows, and vanished seas and freshwater lagoons; and later geologists identified fossil coral reefs, fossil soils with fossil tree stumps embedded in them, and an-

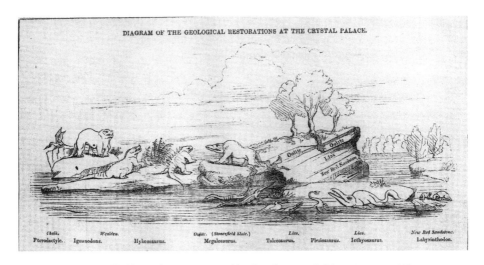

DIAGRAM OF THE GEOLOGICAL RESTORATIONS AT THE CRYSTAL PALACE.

| Chalk. | Wealden. | | Oolite. (Stonesfield Slate.) | Lias. | Lias. | New Red Sandstone. |
| Pterodactyle. | Iguanodons. | Hylæosaurus. | Megalosaurus. | Talæosaurus. | Plesiosaurus. | Icthyosaurus. | Labyrinthodon. |

**FIG. 9.2** The life-sized reconstructions of fossil reptiles, several of them enormous, at the Crystal Palace in suburban London; this famous outdoor exhibit was constructed near the huge steel-and-glass building that housed the relocated Great Exhibition of 1851. The spectacular extinct animals were arranged in sequence, with time flowing here from right to left, from the *Labyrinthodon* of Triassic age (right), through forms such as the *Ichthyosaurus* and *Megalosaurus* of Jurassic age (center), to the *Iguanodon* and *Pterodactyle* of Cretaceous age (left). The artificial cliff behind the reptiles reproduced in miniature the pile of rock formations in which the fossils were found, as evidence for their correct sequence in the Earth's history. This drawing illustrated a lecture given in 1854 by Waterhouse Hawkins, the designer of the models (they are still on show in the public park, though the building is no more). Some of the models (e.g., of the *Iguanodon*) were based on very incomplete fossil material, and differ markedly from modern reconstructions of the same animals.

cient beaches with ripple marks, fossil footprints, and even the marks of raindrops. All such features were grist to Lyell's actualistic mill, as evidence for the steady uniformity of physical processes throughout the Earth's history. But many of these examples of the familiarity of the deep past were in fact discovered by his catastrophist critics such as Buckland, who were no less keen to use comparisons with the present as far as they could be shown to apply.

One important sign of this taming of the deep past was the recognition that quite different kinds of rock might have been deposited during one and the same geological period, just as many different environments exist alongside one another in the present world. In contrast, William Smith's kind of stratigraphy had assumed a unique and unvarying sequence of distinctive formations, each with its own equally unvarying "characteristic fossils": a kind of natural "layer

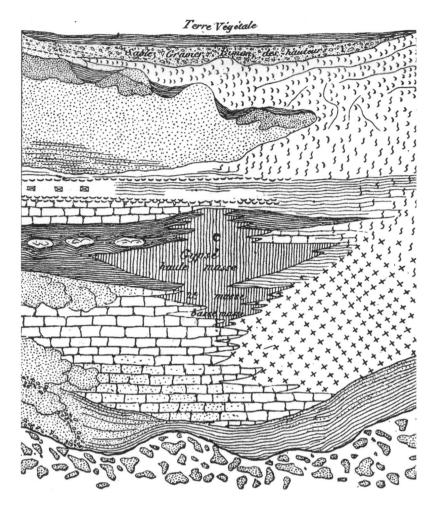

**FIG. 9.3** Constant Prévost's "Theoretical Section of the Parisian Formations" (1835), showing his interpretation of the stratigraphy of the Paris Basin, as established by fieldwork. The vertical dimension shows the sequence of Tertiary strata, and therefore also the passage of time, from the oldest at the base, overlying the Chalk (depicted with scattered flints), to the youngest at the top, underlying the surface soil (*Terre végétale*). The horizontal dimension is that of space, across the Paris Basin from what in the Tertiary period were marine conditions (left) to freshwater conditions (right). The diagram depicts Prévost's claim that, for example, the Coarse Limestone (brick pattern, lower left) and a contrasting sandy formation (pattern of crosses, lower right) had been deposited in different parts of the Paris Basin during much the *same* period of time (they would therefore be defined later as different *facies*). The Gypsum formation (*Gypse*, vertical shading, center), famous for its fossil mammals, was interpreted as the product of a temporary lagoonal environment halfway between marine and freshwater conditions and about halfway through the sequence of events. This kind of diagram was widely adopted by later geologists (and their modern successors) as an effective way to summarize in visual form the often complex variations of strata interpreted in terms of the history, geography, and ecology of their original environments.

cake," to use the metaphor adopted by modern stratigraphers. But Cuvier and Brongniart had found one case of spatial variation that was too prominent to ignore: the Coarse Limestone in some parts of the Paris Basin was replaced by a thick sandstone formation in other parts, both occupying the same position in the pile of rocks. Prévost, their compatriot and critic, later explained this anomaly by reinterpreting the whole sequence as the product of a continually shifting boundary between marine and freshwater conditions across the future Paris region, with the famous gypsum formation as a temporary lagoonal environment situated between them.

This kind of interpretation in terms of local environments was soon extended and given a useful name, when the young Swiss geologist Amanz Gressly mapped the geology of the Jura hills on the Franco-Swiss border. He found that in a specific part of the Jurassic system, but at different localities, there were different sets of rocks and fossils, all clearly of the same age. He called them different *facies* ("faces," in the sense of facial expressions). Gressly interpreted these facies as representing, for example, coral reefs, the shallow lagoons inside the reefs and the deeper water outside. He portrayed their spatial distribution on what would later be called a *palaeo-geographical* map. This new concept of facies signaled the transformation of stratigraphy into a fully *historical* form, in which piles of rock formations were routinely interpreted (as they still are) as the traces of events and environments that have varied in complex ways through both time and space. One large-scale example of the effectiveness of this idea was the recognition that the most puzzling of all the anomalies underlying the "great Devonian controversy" had a quite simple explanation. The Devonian formations were of two contrasting facies formed during one and the same Devonian period, but in two contrasting environments in different parts of Europe, Russia, and North America: the Old Red Sandstone, with its strange early fishes, probably in freshwater; the other Devonian rocks, with their varied molluscs, corals, and so on, under marine conditions.

## GEOLOGY GOES GLOBAL

The Devonian controversy had expanded within a few years from one English county to the whole of northwest Europe and then out to the Urals in Russia and to New York State and beyond in North America. This was just one example of the dramatic extension of the scope of geology as a whole in the course of the 19th century. Geologists' widening knowledge of the sequence of formations and of the whole fossil record was made possible by the rapidly accumulating stock of detailed local surveys of rocks and collections of fossils from all parts of the world. This in turn reflected the rapid pace of exploration—geological surveys usually followed hard on the heels of the geographical—with the expansion of Western global commerce and colonialism, and the increasing exploitation of natural resources of all kinds. Combined with the growing scientific independence of nations beyond Europe, notably Russia and the United States, this began to give geologists the confidence to make generalizations about the Earth's history that might be valid worldwide.

The outstanding example of this globalization of geology was the work of the Austrian geologist Eduard Suess, who tackled the long-standing puzzle of the origin of the Earth's great mountain ranges. Starting in the 1870s with his own and his contemporaries' fieldwork in the Alps, he utilized research by many geologists of many nations to create a worldwide synthesis, which he published in four huge volumes as *The Face of the Earth* (*Das Antlitz der Erde*, 1883–1904). Like most geologists of his time, Suess believed that a mass of evidence supported the model of a slowly cooling Earth. Like Élie de Beaumont earlier in the century, he envisaged the solid crust as having crumpled from time to time to adjust to the steady shrinkage of the deeper parts of the Earth: the shriveling skin of an apple as it dries out was the homely if imperfect analogy that was often used to explain this "*contractionist*" theory. Suess rejected Lyell's steady-state model—which Darwin too had adopted—of vast plates in the Earth's crust, endlessly rising and sinking and in some places rising enough to form mountain ranges. He replaced it with a dramatic model of localized *horizontal* movements, as the Earth's crust contracted by being crumpled along certain specific lines.

The rocks had either been buckled into huge open folds (as for example in the Appalachians, which were being surveyed by American geologists), or they had been pushed right over each other in enormous "*overfolds*," or even *thrust* on top of one another, so that in some places older formations were found overlying younger ones

**FIG. 9.4** A section through a part of the Alps, showing three huge overfolds or *nappes* that have been thrust over one another from the south (right) towards the north, turning some of the rock formations upside-down. The unexpectedly complex structure implied a huge amount of shortening of the Earth's crust in the Alpine region. This section, published in 1902 by the French geologist Maurice Lugéon, was based on his and others' detailed fieldwork in the later 19th century. The formations involved were mostly, by their fossils, of Secondary (Mesozoic) age, so the Alpine mountain-building or *orogeny* must have taken place subsequently, some time in the Tertiary (Cenozoic) era. Combining evidence (as shown here) from several parallel traverses in this deeply eroded mountain region, it was possible to construct a reliable picture of the complex three-dimensional structure. Geologists accepted that these extraordinary movements really had happened during the Earth's history, even though there was no agreement on the causal forces involved.

(as for example in the Alps, which were being unraveled by European geologists). In all such mountain ranges the Earth's crust was, on a vast scale, like a crumpled tablecloth or *nappe* (the French word was later adopted by geologists everywhere to denote these huge displaced masses of rock). Suess recognized that these enormous movements, apparently powered by the crumpling of the solid crust as the Earth cooled and its interior slowly shrank beneath it, had not necessarily been as sudden or violent as Élie de Beaumont had supposed. The *orogenic* ("mountain-building") process might have been imperceptibly slow on the timescale of human lives, but it could still have been catastrophic by the standards of geological time. Suess criticized Lyell for his extreme "quietism" about the pace of geological processes, but his own contractionist theory, no less than Lyell's, took full account of the vast scale of the Earth's history. In effect, many of the earlier arguments between Lyell the uniformitarian and his catastrophist critics had become obsolete.

Suess took the bull by the horns by beginning his great work with a review of the then newly discovered cuneiform record of a Mesopotamian flood. He showed that it—and by implication the biblical version too—were highly plausible, however bizarre some of their recorded details might appear to modern eyes. They were not to be dismissed as merely mythical, because regional catastrophes of this kind had in fact been quite frequent, even within recorded human history. This case became just a geologically recent example in Suess's great synthesis of the deep history of disruptive events of all kinds and on different scales of space and time. Like some of his contemporaries, Suess distinguished three successive orogenies or major phases of mountain building in Europe, widely spaced in time: first the *Caledonian* orogeny (from the Roman name for Scotland) before the Devonian period; then, before the Permian period, an orogeny that was later known as the *Hercynian* (from the Roman name for wooded hills such as Werner's home region of the Erzgebirge in Germany); and most recently the *Alpine* orogeny within the Cenozoic era (the classic fold on the Isle

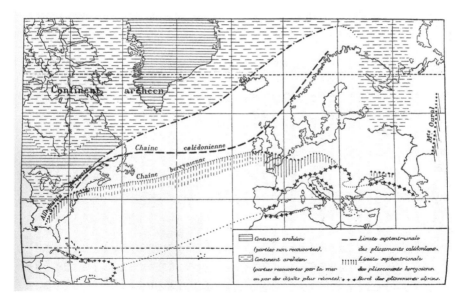

FIG. 9.5 A map of the North Atlantic region, showing the transatlantic correlations that the French geologist Marcel Bertrand suggested in 1887 for the great belts of highly folded rocks— *Caledonian*, *Hercynian*, and *Alpine* (the last shown by lines of crosses)—attributed to three successive *orogenies* or periods of mountain-building movements in the Earth's crust. They were taken to mark the progressive southward growth of an ancient super-continent (horizontal shading). What is now the floor of the Atlantic was assumed to have subsided more recently, separating Europe and Africa from the Americas. All the older mountains (e.g., the Scottish Highlands and the Alleghenies or Appalachians) had been eroded so far that they were no longer high mountains like the Alps.

of Wight being as it were a small outer ripple of this). These could be matched with the lines of similar movements of the same three ages on the other side of the Atlantic, suggesting that the crumpling had been synchronized worldwide. In effect, all this complemented the kind of history provided by stratigraphy and the fossil record, enriching it with a history of the Earth's episodic disruptions. In all respects that history was nothing if not eventful.

Converting stratigraphy into the archives of the Earth's *history* had already come to maturity, as it were, in the work of the English geologist John Phillips who, appropriately, was William Smith's nephew and had been informally his apprentice. Phillips became one of the world's leading paleontologists, making himself familiar with the fossil record in every geological period; again appropriately, he rose to occupy what had been Buckland's position at Oxford. It was Phillips who, for example, had been able to confirm that the marine fossils of what came to be called the Devonian period were intermediate in character between Murchison's Silurian fauna and the Carboniferous fauna on which Phillips himself was an expert, so that they were likely to be of intermediate age. However, he was uneasy about "system" names such as Jurassic, Devonian, and Silurian, on the grounds that they referred to particular regions (the Jura hills, Devonshire, the Welsh Marches) and might not be appropriate to characterize deposits of the same ages elsewhere; he also wanted to puncture Murchison's overweening ambitions for worldwide recognition of "his" systems. Phillips wanted to replace them with terms based on major changes in the long-term *history of life*, which increasingly seemed to be a unique sequence that was valid worldwide.

So in 1841 Phillips had proposed a threefold division of the entire fossil record into a sequence of three great eras in the Earth's global history: a *Palaeozoic* era of "ancient life," a *Mesozoic* era of "middling life," and a *Cenozoic* era of "recent life." These were analogous to the fossil-based periods (*Eocene*, etc.) that Lyell had applied to the Tertiary formations, but they were on a far larger scale. They also had an obvious analogy, which Phillips did not need to make explicit, with the traditional threefold division of human history into Ancient, Medieval, and Modern. Palaeozoic life, with trilobites and tree-ferns, was distinct from Mesozoic life, with ammonites and giant reptiles, which in turn was distinct from Cenozoic life, with varied extinct mammals but other organisms of quite modern appearance. Phillips's three great eras were soon adopted by geologists worldwide (and they remain valuable to their modern successors). His well-informed impression of the fluctuating diversity of life, as reflected in the total fossil record, convinced him that his three eras were not merely arbitrary or accidental but had a real basis in the history of life.

In 1860 Phillips used his presidential address to the Geological Society in London, and then a prestigious public lecture in Cambridge, to argue that the known fossil record, despite being far from perfect, was already complete enough to support an interpretation of the history of *Life on the Earth* (the published lecture's title) as having been broadly "progressive," with the successive introduction of "higher" forms of life, both animals and plants: as he summarized it emphatically, "The earth, then, has a HISTORY." His restatement of this mainstream scientific consensus was in response to Darwin's *Origin* of the previous year, with its Lyellian assumption that the fossil record was

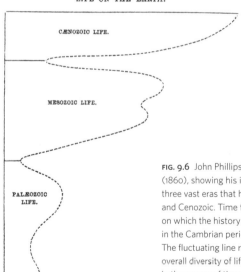

LIFE ON THE EARTH.

CÆNOZOIC LIFE.

MESOZOIC LIFE.

PALÆOZOIC LIFE.

FIG. 9.6 John Phillips's diagram, published in his *Life on the Earth* (1860), showing his interpretation of its total history in terms of the three vast eras that he had earlier defined as Palaeozoic, Mesozoic, and Cenozoic. Time flows upwards, as in the pile of rock formations on which the history was based, from the apparent beginnings of life in the Cambrian period (bottom) to the present world of life (top). The fluctuating line represents Phillips's impression of the changing overall diversity of life in the course of its history: generally increasing in the course of time, but with two prominent troughs indicating that his three eras had a real basis and were not merely arbitrary. Neither dimension could be quantified, but this hardly reduced the value of the diagram, which was based on Phillips's almost unrivaled museum and field experience (his successors will notice its striking similarity to modern and more quantitative diagrams of the same kind, based of course on vastly more extensive information about the fossil record).

extremely fragmentary, and specifically Darwin's claim that it could not be used as evidence against his theory of very slow evolutionary change. Phillips conceded that the striking contrasts between the animals and plants of his three great eras, and the plunges in diversity that separated them, might be due to a relatively imperfect fossil record at those points, but alternatively there might have been at least two episodes of mass extinction.

On this issue, the opinions of geologists during the rest of the 19th century were spread out across a wide spectrum. At one extreme was Lyell who, as the leading and indeed almost sole uniformitarian, continued (until his death in 1875) to explain every appearance of sudden change as the illusory product of an extremely imperfect fossil record. This carried an implicit prediction that further fieldwork in new areas, or the closer study of those already known, might fill in some of the gaps and lessen the apparent discontinuities. To some extent this prediction was fulfilled, as was shown, for example, by the interpolation of new periods (*Paleocene*, *Oligocene*) into Lyell's own scheme for what by then was called the Cenozoic era. But other apparently sudden breaks, particularly those separating Phillips's three great eras, remained obstinately unfilled. This suggested that at least some apparent episodes of mass extinction, particularly those at the ends of the Palaeozoic and Mesozoic eras, might have been genuinely exceptional events in the Earth's history, which would need to be explained and not merely explained away.

This kind of catastrophist interpretation was taken furthest by some French geologists, notably Alcide d'Orbigny, who interpreted every discontinuity in the fossil record as the trace of a sudden "revolution" of some kind. But the number and frequency of such putative events increased to the point where they became rather implausible. In contrast, British geologists were so strongly influenced by Lyell's persuasive argument that they continued to avoid suggesting *any* kind of event in the deep past that might have been more sudden, intense, or violent than those recorded in the present world. They were swayed by Lyell's forceful claim that to postulate catastrophes of any kind was deeply unscientific, although this entailed dismissing all the scientifically legitimate explanations (mega-earthquakes, mega-tsunamis, mega-eruptions, etc.) suggested by those who argued for the historical reality of such natural events.

In fact, *all* geologists in the 19th century, whatever their opinions on the reality of occasional catastrophic events in the deep past, used the actualistic maxim—"the present is the key to the past"—as a matter of course. Indeed the method was extended, in that the geologically recent but prehistoric past could be taken as a guide to what might have happened in the still deeper past, which was often much more obscure. So, for example, the increasingly well-understood signs of a widespread Ice Age in the Pleistocene period were used, by analogy, to identify the unexpected traces of another and far earlier glacial period (which incidentally made the later one less of an anomaly, if in fact it had not been unique). The distinctive till or boulder clay that had formed under Pleistocene ice sheets could be matched with far older deposits, the solid rocks later called "*tillites*," which were similarly full of angular boulders of all sizes. These tillites were discovered in formations of late Palaeozoic (Carboniferous or Permian) age. But they were not found in Europe or North America, where rocks of this age included the Coal formation with its fossil forests of tropical appearance, and also sandstones and salt deposits that looked like the relics of hot deserts. Instead these tillites, which were taken to be traces of ice sheets formed during an ancient Ice Age, were found in Australia, in southern Africa, and, most puzzlingly, in India.

This was another product of the steady globalization of geological exploration. It suggested a distribution of climates strikingly different from that of the present world, yet it also failed to fit what might have been expected on a steadily cooling Earth. Geologists who were surveying India—those directing the Survey were British, though many of those doing the fieldwork were Indians—described formations that they called collectively the "*Gondwána* system." By the usual stratigraphical criteria these rocks were clearly of late Palaeozoic age, but they were more like those being explored in southern Africa and Australia than those already well known in Europe and North America. In the 1870s the geologists in India suggested that Africa, Australia, and India had once been parts of a single huge landmass. This startling idea gained wider acceptance after it was endorsed by Suess, who named the putative super-continent "*Gondwana-Land*"; and one of the geologists working in India later suggested enlarging it to include South America and even Antarctica. Apart from the latter (the geology of which was as yet almost unknown), these widely dispersed landmasses had in common not only the occurrence of ancient tillites but also some distinctive fossils. These included some distinctive early reptiles, and also the plant *Glossopteris*, which in effect replaced the well-known fossil plants of about the same age in the Coal formation in Europe and North America. This idea of a former super-continent mostly in the Southern Hemisphere was also supported by a growing body of evidence (later called *biogeographical*) that the same landmasses also had many distinctive *living* animals and plants in common. How all these terrestrial organisms, fossil and living, had come to be so widely distributed remained controversial; but it was commonly suggested that *"land-bridges"*—somewhat like the present isthmus of Panama, but perhaps much broader—might once have connected continents that are now separated by vast spans of ocean. Whatever the causal explanation, the Earth's ancient geography and climates were evidently as strange and unfamiliar as its ancient trilobites and dinosaurs. These were just some of the ways in which, in the later 19th century, the history of the Earth and its life became more unexpected and eventful than ever, in part because the problems and the evidence were now global in extent.

## TOWARDS THE ORIGIN OF LIFE

One of the most intriguing aspects of the fossil record, as it became known in increasing detail in the later 19th century, was the puzzling question of its very beginning at the start of the Palaeozoic era. This had an obvious bearing on the fundamental problem of the origin of life itself. In the 1850s it became clear that the lowest and oldest of the Palaeozoic systems, the Cambrian, which Sedgwick had based on rocks in Wales with very few fossils of any kind, was chararacterized elsewhere by quite abundant fossils that were distinct from those in Murchison's overlying (and therefore later) Silurian. The formations in which these Cambrian fossils were found were for a time called *"Primordial,"* because their fossils were the earliest unambiguous signs of life of any kind. But they were not the small and simple organisms that were expected if either Lamarck's or Darwin's kind of evolution was on the right lines. They were complex and highly diverse, and some were large in size. Even if they were only a modest foretaste of the much greater diversity still to come, the Cambrian seas had evidently

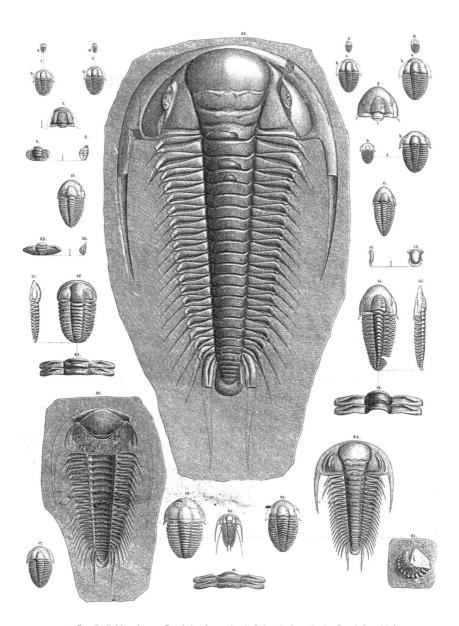

**FIG. 9.7** Fossil trilobites from a Cambrian formation in Bohemia (now in the Czech Republic), underlying and therefore older than the Silurian fauna. These superbly preserved fossils were described in the 1850s by the French emigré civil engineer Joachim Barrande. The largest specimen shown here was 18 cm (7 in) in length and others were much larger; the smaller ones here are of different species but equally complex in structure. A full range of the growth stages of each species was preserved. Barrande called this fauna *"Primordial"*; it was later recognized as equivalent to Sedgwick's "Cambrian." The still older pre-Cambrian (or Precambrian ) rocks had *no* clear fossil record, which was a major problem for Darwin's kind of evolutionary interpretation of the history of life.

teemed with Cambrian life. It seemed to be at the very start of the fossil record, yet it did not look obviously "primitive."

This was all the more puzzling, because the Cambrian formations were usually found directly overlying a "basement" composed of the rocks formerly called "Primary," which were now usually termed *Archean* ("ancient" or "primitive"). Their mode of origin was controversial, but most of them were crystalline (granite, gneiss, schist, etc.) and were agreed to be extremely unlikely to contain any fossils at all; certainly none was found. Since they clearly preceded the Palaeozoic era, they were often attributed to an earliest *Azoic* ("no life") era in the Earth's history, perhaps a time when the Earth's surface was still too hot to support life of any kind. An alternative explanation, which had some support among geologists, was Lyell's metamorphic interpretation of many of these same rocks: that while buried in the depths of the Earth they had been so severely altered by its intense internal heat that all traces of their fossils had been destroyed. Either way, there seemed to be little hope of ever discovering whether any life had existed during the pre-Cambrian (or Precambrian) history of the Earth, or what it had been like.

However, geologists were not so easily defeated. In some regions formations with Cambrian fossils were found to be underlain not by such hopeless "basement" rocks but by quite ordinary-looking sandstones and shales: if they had been Palaeozoic in age (if, say, they had *overlain* Cambrian formations), they would have been considered hopeful as possible sources of fossils. So these Precambrian rocks were assiduously searched for fossils, in the hope of finding the ancestors of the Cambrian fauna: they were optimistically attributed in advance to a *Proterozoic* ("earliest life") era. When, for example, a possibly organic structure was found in a Precambrian formation in Quebec, the Canadian geologist John Dawson named it *Eozoon* ("dawn-life") and publicized it in his *Life's Dawn on Earth* (1875). Since it was quite different from the Cambrian organisms, and unlikely to have been the ancestor of any of them, Dawson hoped it would help to refute Darwin's kind of evolution, to which he was strongly opposed. But *Eozoon* was controversial from the start, and by the end of the century it had been dismissed as wholly inorganic and merely a product of metamorphic alteration.

Despite this setback, preliminary surveying of Precambrian for-

**FIG. 9.8** The supposed organism *Eozoon canadense* from a Precambrian limestone in Quebec, as illustrated by John Dawson in his *Life's Dawn on Earth* (1875). This drawing is of a thin transparent slice through the solid structure, as seen through a microscope (the then new technique of making such "thin sections" was dramatically improving knowledge of rocks and fossils of all kinds). Dawson interpreted the structure as a giant protozoan, and reconstructed it with inferred "animal matter" (dark areas) and a preserved "calcareous skeleton" (light areas). He claimed that it provided evidence of life long before the diverse fossil animals in Cambrian rocks. But it was not a plausible evolutionary ancestor for any of them, and it was later dismissed as a purely inorganic product of metamorphic alteration of the rock. This left the vast spans of Precambrian time with no clear fossil record at all.

mations continued to encourage the search for signs of Precambrian life. When, for example, the young geologist Charles Walcott, having joined the new United States Geological Survey, was exploring and surveying the western states, he found structures that were later named *Cryptozoon* ("hidden-life"), referring appropriately to their enigmatic character. They were large pillow-like mounds, which did look possibly organic in origin; but what kind of organism had formed them remained a mystery. Whatever it was, it did not look a likely ancestor for any of the much more complex Cambrian fossils. Nor did the very rare and very dubious scraps of other alleged Precambrian fossils found elsewhere. The Precambrian formations were therefore still left without any unambiguous fossil record.

This was a serious problem for evolutionary theorizing, at least in the Darwinian form that demanded extremely slow and gradual change. If all the highly diverse Cambrian organisms had evolved slowly and gradually from "some one primordial form," as Darwin had conjectured, the Precambrian history of life would have had to be almost unimaginably lengthy: extending at least as far back before the Cambrian as the total time that had elapsed since the start of that period. But if so, the first half—or more—of the total history of life was apparently without any clear fossil record at all. There were two possible explanations. There might be no fossil record because there had been no life to be recorded: all the Precambrian might deserve to be called Azoic, not Proterozoic. In that case, and if the diverse Cambrian organisms had indeed evolved from "some one primordial form," this would have had to happen relatively quickly in what would later be termed a *"Cambrian explosion"* or rapid diversification of life around the start of the Cambrian period. This would be a decidedly un-Darwinian kind of evolution. Alternatively, if Precambrian time had in fact been long enough to allow for the slow Darwinian evolution of the Cambrian fauna, the lack of any record of all that evolutionary development became even more of a problem. To suppose that all the ancestors of the diverse Cambrian animals had been "soft-bodied," and therefore unlikely to be preserved as fossils, and that they had all acquired "hard parts" (such as shells that could easily be fossilized) at about the same time, looked like a desperate attempt by Darwinians to extricate themselves from a tight corner. At the end of the century the origin and early evolution of life, before the Cambrian period, remained deeply mysterious.

## THE TIMESCALE OF THE EARTH'S HISTORY

It should be clear, from the narrative in this and previous chapters, that geologists in the 19th century made spectacularly successful progress in reconstructing the Earth's history, without benefit of even an approximately quantitative or *"absolute"* timescale. In practice it hardly mattered to them that the history of the Earth and its life could not be quantified. The increasingly well-established *"relative"* timescale of eras and their component periods, based on the observable sequence of rock formations, was more than enough to keep geologists busy and productive. All that mattered to them in practice was that, as Scrope had memorably put it, all the evidence shouted out "Time!—Time!—Time!"; or, as Conybeare had insisted, Lyell could have as many "quadrillions of years" as he liked, provided he could show that such an order of magnitude was really needed. Contrary to persistent later myths and misconceptions about this issue, all geologists in the 19th century—mainstream catastrophists no less than Lyell the uniformitarian and his lone disciple Darwin—were convinced that the magnitude ran at the very least into millions of years; but how many millions, or tens or hundreds of millions or perhaps even billions, made little if any practical difference to their impressively fruitful reconstruction of the deep past. Although Lyell's eloquent rhetoric often suggested otherwise, even the most ardent catastrophists did not invoke catastrophes because they felt squeezed for time, but because they believed that the evidence demanded interpretation in terms of unusually sudden and even violent events, however vast the total length of time within which these events had occurred.

Nevertheless, the advantages of being able to turn the qualitative timescale of successive eras and periods into a quantitative one—calibrated, however approximately, in years—became ever more obvious, the more securely the sequence of geological events was established. Without such a timescale, geologists were in the position that historians would have been in, if they had had reliable evidence about the correct sequence of distinctive periods in European history (Mediaeval, Renaissance, Enlightenment, etc.) but no firm dates and therefore no clear idea whether particular key events had been separated by several centuries or only a few decades. Just as the scholarly chronologers of the 17th century had tried to fix accurate dates back to the start of human history, so a quantitative or absolute timescale for geology would provide the Earth with its own similar chronology, or what at the end of the century was named "*geochronology*."

Deluc's concept of "nature's chronometers," such as the recorded rates of growth of deltas, had not been forgotten (though he himself had been). Deluc had wanted to derive an approximate date for a single decisive event, the physical "revolution" that he equated with the biblical Flood. But in the 19th century the challenge for geologists was more daunting: to estimate the *total* time represented by the rocks and fossils they were describing, at least to the right order of magnitude. Lyell had hoped to do this when he devised a natural chronometer

Geological Scale of Time.

| | Periods. | Systems. | Life. |
|---|---|---|---|
| | Cænozoic. | Pleistocene. | Man. |
| | | Pleiocene. | Placental Mammals. |
| | | Meiocene. | |
| | | Eocene. | |
| | Mesozoic. | Cretaceous. | |
| | | Oolitic. | Marsupial Mammals. |
| | | Triassic. | |
| | Palæozoic. | Permian. | Reptiles. |
| | | Carboniferous. | |
| | | Devonian. | Land Plants. Fishes. |
| | | Siluro-Cambrian. | Monomy. Echinod. Pterop. Heterop. Dimy. Gasterop. Annel. Polyzoa. Zooph. Brach. Crust. |

FIG. 9.9 John Phillips's "Geological Scale of Time," as published in his *Life on the Earth* (1860); it embodied an implicit consensus among most geologists at this time. The pile of rock forma-tions, in their stratigraphical sequence as "*Systems*" (middle column), was divided into the three vast eras or "*Periods*" of time (column on left) that Phillips had earlier defined as Palaeozoic, Mesozoic, and Cenozoic. The history of "*Life*" (column on right) showed the points at which various major groups of organisms made their *first* appearance in the fossil record; the abbrevi-ated names at the bottom refer to the highly varied *invertebrate* or backbone-less animals found in the Cambrian and Silurian strata at or near the start of the fossil record. This diagram also depicted Phillips's provisional attempt to quantify the timescale of life's total history, based on the maximum known thicknesses of the formations assigned to each "system." The whole span of time from the start of the Palaeozoic era (bottom) to the present (top) was divided into ten arbitrary units of equal duration, numbered both in units before the present (10 to 1, outside left margin) and in units since the start (1 to 10, inside left margin); this was analogous—probably unknowingly—to the early chronologists' use of timescales both of "Years Before Christ" (BC), counting backwards, and of "Years of the World" (*Anni Mundi*, AM), counting forwards. Phillips remained uncertain how his quantitative scale of time should be calibrated, though like all geologists in the 19th century he had no doubt that it should be estimated *at least* in many tens of millions of years. Since the diagram represented the history of *life*, the Precambrian, with no known fossils, was omitted altogether.

by plotting the changing percentages of extinct and extant species of molluscs in the Tertiary formations; he had tried to calibrate this in years and had hoped to extend it to the Earth's earlier history. A more ambitious attempt at geochronology was later made by Phillips, who was able to deploy his outstandingly wide knowledge of the entire pile of formations, from Cambrian through to Pleistocene, and of the worldwide fossil record they contained. In the 1850s he combined what little was known about the rate at which new sediments are accumulating at the present day, for example in the delta of the Ganges, with the reported maximum thicknesses of strata in the successive stratigraphical systems; as usual, the present was being taken as the key to the past. But there were so many uncertainties, and so many assumptions were required, that Phillips was reluctant to publish any estimate. Darwin's *Origin* forced his hand.

Darwin, in his only shot at tackling this issue, estimated about 300 millions of years (or 300 Ma, in modern geologists' shorthand) for the slow erosion of the shallow dome of Chalk and other Cretaceous formations in the Weald region of southeast England (bounded by the distinctive Chalk hills of the North and South Downs, and visible from near Darwin's rural home on the former). This was roughly equivalent to the Cenozoic era, the time since the deposition of the Chalk. In 1860 Phillips used the prestigious occasions already mentioned to give his reasons for rejecting Darwin's figure as a gross overestimate. Darwin tacitly accepted the criticism, for he promptly dropped his calculation from all further editions of the *Origin* (it was, incidentally, far *in excess of* the modern radiometric date). Phillips's own tentative estimate was of about 96 Ma for the whole span of time from the start of the Cambrian period (and of the Palaeozoic era) to the present, though he later reduced this somewhat, and anyway he knew that any figure was little better than a shot in the dark.

Most other geologists, apart from Lyell and Darwin, regarded Phillips's estimate as reasonable, and as reliable as any could be. Significantly it was Phillips, representing their implicit consensus, who was consulted when in 1861 the leading Scottish physicist William Thomson wanted to compare the opinions of geologists on the likely magnitude of the Earth's timescale with his own estimate derived from the quite independent evidence of physics and cosmology. (Near the end of his life, in recognition of his important role in developing transatlantic telegraphy, Thomson was ennobled as Lord Kelvin, the name by which he has been known ever since and the one that is convenient to use here.) Kelvin told Phillips that his own preliminary calculations suggested a total age of the Earth somewhere between 200 and 1000 Ma, a range with which Phillips's estimate of 96 Ma (just since the start of the Cambrian period) was more or less compatible. Kelvin's figures were based on his cosmological theory about the origin of the Solar System, and specifically that the Sun was continually using up its initial source of heat. According to the then recently formulated laws of thermodynamics, the Sun would necessarily have a finite life span, which would limit the possible age of the Earth. Kelvin's theory also incorporated the geologists' standard model of a slowly cooling Earth and their evidence for its internal heat. In 1863, after reworking his calculations, Kelvin published a revised range between 20 and 400 Ma for the total age of the Earth; if further assumptions were made, he could narrow this down to a likely figure of 98 Ma. Allowing for all the uncertainties, this was still just about compatible with Phillips's almost identical figure (for post-Precambrian time alone). Anyway Phillips was satisfied. The following year he commended Kelvin's estimate to his fellow geologists and welcomed the physicist's contribution to the scientific understanding of the Earth. He later collaborated with Kelvin in developing instrumentation to improve measurements of temperature in mines, which could give valuable evidence about the Earth's internal heat. There was certainly no deep conflict here between geologists and physicists.

In 1866, however, Kelvin delivered a lecture entitled provocatively "The Doctrine of Uniformity in Geology Briefly Refuted," and in further lectures he expanded on this theme. He vehemently attacked those geologists who were ignoring his work and continuing to invoke unlimited amounts of time for the Earth's history, far beyond what he believed the science of physics allowed. His targets were Darwin and Lyell. These two were indeed prominent figures, but hardly representative of geologists as a whole; surprisingly, Kelvin did not use Phillips to support his own position. In fact, both the content and the timing of his attack on "uniformity" suggest that his main target was evolution, or rather Darwin's specific brand of evolution. This was because it demanded vast amounts of time for the extremely slow process of natural selection to have its effects; and natural selection

undermined the sense of nature's designfulness, which was as important to Kelvin as it was to many of his contemporaries. During the rest
of the century—and even after his uniformitarian targets had died—
Kelvin continued his campaign against "uniformity." To the dismay of
geologists he claimed that on refining his calculations and assumptions a further reduction of his estimates was required. In 1881 he reduced even the upper limit for the likely age of the Earth to 50 Ma; in
1897 he reduced this again to 40 Ma, while giving 24 Ma as his best
possible estimate. Although some physicists were sceptical about his
figures and the assumptions on which they were based, others supported him: one argued for a still greater reduction, to less than 10 Ma,
which made even Kelvin look quite moderate.

Meanwhile most geologists would have been content if the physicists' timescale had remained somewhere near Phillips's earlier estimate, that is at about 100 Ma (for post-Precambrian time). At the
very end of the century this order of magnitude was supported from a
different direction, on grounds almost as independent of mainstream
geology as Kelvin's. The Irish geologist John Joly, who was also a more
than competent physicist, revived and updated Halley's far earlier suggestion that a natural chronometer could be derived from the rate at
which the world's rivers are currently adding to the salt content of its
oceans. Joly estimated 90–100 Ma for the age of the oceans, which
was taken to be the time since the Earth's primitive crust cooled sufficiently to allow water to condense on its surface.

As the 19th century ended, then, there was a widening gulf between geologists and physicists on this issue of the Earth's total age.
The stand-off was not helped by what geologists regarded as the sheer
arrogance of the physicists who—not for the last time—openly treated
other sciences as far inferior to their own. The age of the Earth was just
one of the many problems about the history of the Earth and its life
that, as sketched in this chapter, were still frustratingly obscure and
controversial. The next chapter moves into the 20th century and to
their partial resolution.

# 10

# Global Histories of the Earth

## DATING THE EARTH'S HISTORY

Around the start of the 20th century, dramatic developments within physics itself knocked the bottom out of Kelvin's estimates of the total age of the Earth. Many physicists had believed that their science was almost complete, with only minor refinements still to be made, but this assumption was blown apart by a series of fundamental new discoveries. Among these was the detection of previously unknown kinds of radiation, the first being appropriately named "*X-rays*." For geologists, however, the crucial new observation, made by the French physicist Pierre Curie in 1903, was that heat was being continuously generated during what his Polish wife and collaborator Marie had already described as "*radio-activity*." This strange and utterly unexpected process was found to be occurring naturally in certain kinds of rock. It was therefore clear that the Earth's internal heat could not be, or not wholly, a relic of its incandescent origin. If heat was being generated from a previously unrecognized source, all estimates of the age of the Earth based on its supposed rate of cooling were almost worthless; at best, they would represent a minimal age, and the real age might be much greater. In 1905 the physicist Ernest Rutherford, a New Zealander working in

England, used the measurable rate at which Marie Curie's new element *radium* decays into helium—a new "natural chronometer"—to estimate the age of one radioactive mineral sample at 500 Ma; his English colleague Lord Rayleigh then estimated another at no less than 2400 Ma. Their method was soon recognized as being flawed, not least because helium, being a gas, was likely to have leaked out over time. But their results seemed not to be a flash in the pan, because the American physicist Bertram Boltwood derived comparable figures, using the potentially more reliable decay of uranium into lead. These first "*radiometric dates*" (as they were later called), however uncertain, opened up at least the possibility of a total age for the Earth much greater than had previously been imagined, either by most physicists or by most geologists.

One recruit to this exciting and highly international field of physics was Arthur Holmes, a young Englishman who, unlike most other physicists, had also been trained in geology. In 1910 he began to work with Boltwood's uranium-lead method. One of his first dates, of 370 Ma, was for a mineral in a Norwegian rock that in geological terms was Devonian in age. Even this date, from the middle of the Palaeozoic era, was more than ten times Kelvin's final estimate for the *total* age of the Earth (though proportionately lower than Darwin's rash early estimate). Holmes's mentor the English physicist Frederick Soddy then discovered that many elements exist in different "*isotopes*." Together they traced the complex paths by which radioactive isotopes decay, one into another. This enabled them to develop much improved methods for estimating the ages of radioactive minerals. In *The Age of the Earth* (1913) Holmes set out the new case for a greatly extended timescale. He dated his oldest specimens at about 1500 Ma and concluded that the Earth itself could not be less than 1600 Ma in age. The laboratory methods involved were technically difficult, slow, and laborious; and these "absolute" radiometric dates were often difficult to correlate reliably with "relative" geological dates for the same rocks. But the experimental methods and the physical evidence on which they were based—for example the measured rates of decay of the relevant isotopes—became progressively more accurate and consistent.

Geologists had become accustomed to thinking in terms of a total timescale of around 100 Ma, and had been under intense pressure from Kelvin and other physicists to accept an even more constricted

GLOBAL HISTORIES OF THE EARTH [ 237 ]

one. Many were therefore sceptical about its dramatic proposed enlargement. In effect, a common reaction to the physicists' startling
change of mind was "Once bitten, twice shy." But after the First World
War, which had almost brought non-military scientific research to a
standstill, geologists recognized the need to debate the issue afresh.
Two conferences, in Edinburgh in 1921 and in Philadelphia in 1922,
brought together geologists, physicists, astronomers, and biologists.
Some of these diverse *"scientists"*—the umbrella word was coming
into common use around this time, almost a century after it had first
been proposed—remained sceptical; Joly was a conspicuous example.
But most of them agreed that the new radiometric dates, however provisional, were likely to be of the right order of magnitude. Rayleigh,
as a physicist, conjectured that conditions on Earth might have been
suitable for life for anything up to a few billion years. The Oxford geologist William Sollas (who had edited Suess's great work in English)
summarized the surprising new situation: "the geologist, who before
had been bankrupt in time, now finds himself suddenly transformed
into a capitalist with more millions in the bank than he knows how
to dispose of." The echo of Scrope's famous banking metaphor a century earlier—that geology "forces us to make almost unlimited drafts
upon antiquity"—was unmistakeable. A few years later Holmes, now
recognized as a leading expert, gave a committee in the United States
his best estimate for the age of the Earth as not less than 1460 Ma
but probably not more than 3000 Ma (or three billion years). By 1953
the American chemist Clair Paterson, using far more radiometric evidence, was able to derive a greatly improved estimate of about 4.5 billion years. His figure is one that the relevant experts continued to find
reliable, through the rest of the 20th and into the 21st century.

However, while physicists and astronomers were most interested in
this *total* age of the Earth, as a step towards understanding the physics of the Sun and the place of the Solar System in the wider cosmos,
geologists were much more concerned to use radiometric dating to
quantify the sequence of events *within the Earth's history*. They wanted
to know how many of their new-found millions to allocate to each
period; they wanted to put figures, however approximate, to the kind
of timescale that Phillips (and others since his time) had derived
from the relative thicknesses of the sediments that had been deposited in successive periods. Holmes had made an early start with this

by reporting, for example, 340 Ma for a rock of Carboniferous age, 370 Ma for a Devonian one and 430 Ma for a Silurian one: all in the right order. After the First World War, geologists' confidence in radiometric dating was progressively strengthened when further figures continued in the same way to match, ever more consistently and precisely, the relative ages already established by the well-tried methods of stratigraphy. Within initially wide margins of uncertainty, radiometric dates did not contradict geological ones. Those margins were steadily reduced when the new "mass spectrometer" enabled mineral specimens to be analyzed with unprecedented precision. After the Second World War this gradually transformed radiometric dating into a routine procedure that became ever more accurate (and cheaper). Several independent methods, based on the rates of decay of different series of isotopes, proved consistent with each other; and one method or another was found effective for different parts of the Earth's total history. By about the 1970s the dating had become so reliable and precise that many *"Earth scientists"*—another umbrella term that was coming into use, particularly to include both geologists and geophysicists—began to use numerical dates routinely, even giving them priority over their named stratigraphical equivalents. For example, a certain episode of apparent mass extinction might be referred to as an event that happened at 65 Ma in the past, alluding in fact to the end of the Cretaceous period and of the Mesozoic era: just as historians, while debating, say, the impact of Darwin's *Origin*, might refer to the decisive year 1859 rather than to the mid-Victorian period.

By the end of the 20th century radiometric dating had become a powerful and indispensable tool in geology. It was of course dependent on the physicists' assumption that rates of radioactive decay, as measured in the laboratory, had remained constant throughout deep time. But two methods that were independent of any such assumption confirmed—at least for the recent history of the Earth—that the geologists' earlier hunches about the order of magnitude of its timescale had indeed been about right. In the late 19th century the Swedish geologist Gerard de Geer had noticed that sediments deposited in former glacial lakes in Sweden were finely banded in a way that closely resembled the annual rings in the trunks of trees: these were already being used by archaeologists to give accurate dates, for example to wooden beams in old buildings, on a *"dendrochronology"* spanning the centuries of recent human history. De Geer inferred that the thin layers of lake sediment, which he named "*varves*" (from Swedish *varv*), were indeed an analogous record of the annual cycle of the seasons. Varves could therefore be used to construct a similar chronology, calibrated in years, for the stages by which the vast Scandinavian ice sheet had gradually shrunk at the end of the Pleistocene glaciations. In a heroic feat of laborious fieldwork, de Geer and his students matched overlapping distinctive sequences of varves from one former Swedish lake to another, right through to the known point in human history (in 1796) when the last of these lakes had been drained. The whole sequence could therefore be dated accurately, far back into years BC. De Geer's reconstruction of the geologically recent history of Scandinavia first became widely known when he described it at an International Geological Congress in 1910; near the end of his life he published it fully in his monumental *Geochronologica Suecica* (1940). This accurate geochronology based on varves was later confirmed by "*radio-carbon*" dating (the method of radiometric dating most accurate for the most recent part of the Earth's history). Varves provided an accurate "natural chronometer" that would have delighted Deluc: just as the 18th-century savant had estimated it on other grounds, it put the start of the "modern world" only a few millennia in the past.

It was also matched in the late 20th century by a similar chronology, likewise calibrated in years up to the present, preserved in the great ice sheets of Greenland and Antarctica. Cores retrieved from drilling through the ice revealed a layering similar to varves (and tree rings), where the ice had been compacted from the annual snowfall. These annual layers of snow, turned into ice, had even trapped samples of air, from which the changing composition of the Earth's atmosphere could be traced, and traces of dust, from which distant volcanic eruptions could be identified and dated. Ice cores confirmed on a global scale what de Geer's varves had already shown for one specific region: as geologists had long inferred on other grounds, the last Pleistocene glaciation had ended (outside the polar regions) several thousands of years ago. This in turn implied that all the rest of the Pleistocene period, and the rest of the Earth's far lengthier preceding history, must certainly have occupied the vast magnitudes of time that geologists already had every reason to take for granted. The millions in their metaphorical bank account were certainly not worthless

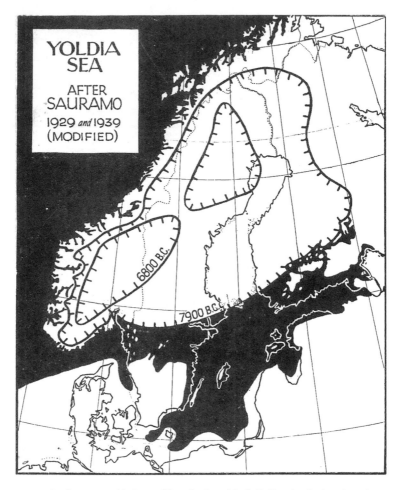

**FIG. 10.1** A palaeo-geographical map of Scandinavia and the Baltic Sea, showing two stages in the shrinkage of the great Scandinavian ice sheet at the end of the most recent of the Pleistocene glacial periods. It was reconstructed *and dated* (in years BC) by Gerard de Geer and his colleagues and students, and based on their detailed fieldwork on the sequence of annual *varves* in the sediments deposited in temporary lakes at the edges of the ice sheet. At the earlier of the stages shown here, at 7900 BC, the Baltic—with its northern part (the Gulf of Bothnia) still covered in ice—was characterized by the mollusc *Yoldia*, which now lives only in much colder Arctic waters. The later stage shown here, at 6800 BC, marked the point at which the shrinking ice sheet split into two smaller ice caps (of which the present Norwegian glaciers are the tiny remnants). This dated reconstruction linked the geologically recent history of the region with the "Mesolithic" phase of human prehistory. It confirmed what geologists had long concluded on other grounds, though with less precision: if even the end of the geologically recent Ice Age was so far in the past by the standards of human history, the whole of the preceding history of the Earth must have extended over spans of time so vast as to be almost beyond human comprehension. This particular map was published in Frederick Zeuner's *Dating the Past* (1946), an influential review of all the methods of "*geochronology*" that were available by then to geologists and archaeologists.

or illusory. They could be invested profitably in reconstructions, and even explanations, of the Earth's immensely long and unexpectedly complex history.

## CONTINENTS AND OCEANS

De Geer's meticulous fieldwork established not only the chronology of the melting of the Scandinavian ice sheet but also the reality of the simultaneous uplift of the land, as shown by the "*raised beaches*" around the shores of the Baltic Sea and the well-known evidence that the land was still rising there (marks carved and dated on rocks at the water's edge since the 18th century recorded how the local sea level was falling and the coastal waters were becoming shallower). Geologists interpreted this uplift of the region as the continuing slow rebound of the Earth's crust, after the removal of the enormous weight of the thick Scandinavian ice sheet at the end of the Pleistocene period. This in turn contributed to a lively debate, which had started back in the 19th century, about the physics of the Earth's crust and the nature of its unseen deep interior. These important arguments about what became known as geophysics must be summarized here very briefly and baldly, in order to concentrate on their implications for the Earth's *history*.

The first major scientific voyages, notably that of the British naval ship *Challenger* (1872–76), had for the first time systematically plumbed the depths of the world's oceans. They had discovered an apparently fundamental distinction between continents and oceans (in this context the continents included the "*continental shelves*" extending beyond the present coastlines, covered by relatively shallow water in contrast to the far deeper oceans). The physical puzzle was how, given the force of gravity, the continental parts of the Earth's surface were able to remain at a much higher average elevation than the oceanic parts. In the late 19th century some geophysicists had suggested that the continents were in effect floating on a slightly denser layer that came to the surface beneath the oceans. Any such "floating" was somewhat metaphorical, for at about the same time the study of earthquakes, using a newly established global network of sensitive instruments, was leading "*seismologists*" to infer that the Earth must be solid throughout, apart from a relatively small liquid core. However, the American geologist Clarence Dutton had argued that the relevant analogy was not with icebergs floating on polar seas but with the ice sheets from which they originate: ice is of course solid and brittle when struck by an ice axe, yet on a much larger scale and over a longer time the ice of Greenland and Antarctica flows slowly down to the sea in great glaciers. In the same way the rocks of what was called the "*mantle*" underlying both continents and oceans, although solid in any everyday sense, might not be rigid on a geological timescale but might act like an extremely viscous fluid.

Dutton had therefore proposed the term "*isostasy*" ("equal standing") to denote what he argued was the buoyancy that in the long run usually keeps continents high above oceans, and mountain ranges higher still. The rebound of Scandinavia then became an example, on a relatively small scale, of this effect of isostatic buoyancy. The importance of isostasy for the Earth's *history* was that, if valid, it almost ruled out the possibility of temporary land-bridges between continents, which had been an attractive explanation of the strange distribution of terrestrial animals and plants. With isostasy it would be physically impossible for portions of the Earth's crust to have risen and fallen so dramatically, being transformed from continent into ocean or vice versa, in the course of the Earth's history. As one geologist put it, "once a continent, always a continent; once an ocean, always an ocean."

This was one of the ways in which, in the early 20th century, Suess's great geological synthesis began to seem inadequate. The idea of crustal plates oscillating slowly up and down—an attractive image ever since the time of Lyell—became difficult to sustain. So did the theory of a slowly contracting Earth, which had failed to account for the conspicuous fact that major mountain ranges are not distributed uniformly around the globe but straddle it along certain specific lines (for example, all down the west side of the Americas from the Rockies to the Andes). Furthermore, detailed fieldwork in the Alps now showed that the huge folding and overthrusting of the rocks—if one reconstructed the tablecloth or *nappe* as it had been before it was crumpled—implied too much crustal shortening to be due to the shrinking of a cooling Earth. In any case the cooling theory itself had been fatally undermined by the discovery of radioactive heat. Altogether, the homely analogy of an apple with a shrivelling skin now seemed misleading. Geologists therefore began to give more attention to the possibility that mountains might have been formed by the *horizontal* movement and compression of adjacent segments of the Earth's crust. The Alps might have been crumpled and forced up into a mountain range when, for some reason, the crust underlying Africa and the Mediterranean region pressed against the rest of Europe; the Himalaya (and the high Tibetan plateau behind it) might have been forced up when India pressed against the rest of Asia.

This idea was startling enough; but even more so was the possibility that these continental masses had not just been squashed together but had in effect collided, having already *moved substantially* in their relative positions on the globe. Yet this might explain the strange facts about what Suess had called Gondwanaland, if that hypothetical former super-continent—comprising India, Africa, South America, Australia, and Antarctica—had somehow split up and the fragments had moved apart. If so, the distribution of certain terrestrial plants and animals might straddle more than one continent, without needing the help of any hypothetical land-bridges. But the problem lay in the "somehow." It was difficult to imagine how any such movement of continents could have been physically powered, even if it was conceded that with "millions" in the bank of deep time the solid globe might have acted as if the rocks were viscous fluids. Not for the first or the last time in the Earth sciences, there was a potential conflict between evidence that something *had in fact* happened and evidence for *how* it could have happened: between history and physics. In earlier cases, notably that of the Ice Age, the relevant scientists had all accepted the historical reality of the glaciations even though the physical or astronomical causes underlying those events remained deeply uncertain; likewise the reality of the spectacular Alpine folding was accepted although there was no agreement about how it had been caused. In this latest case, however, the possibility of "*mobilism*" (as it was later called) was fiercely opposed by those who saw no adequate mechanism for any such wholesale movement of the continents and who therefore declined to abandon their "*fixism*."

The idea of mobilism had already been suggested, but it got little attention until, during the First World War, the German geologist and meteorologist Alfred Wegener published *The Origin of Continents and Oceans* (*Die Entstehung der Kontinente und Ozeane*, 1915). Soon after the war a revised edition was translated into English and other languages, and Wegener's theory of the large-scale *"displacement"* of continents became a hot topic of international debate among Earth scientists. In Wegener's opinion the geological evidence was paramount. Nothing less than large-scale displacement could account for the close similarities between widely separated landmasses such as those attributed to Gondwanaland. In particular, Africa and South America (defined by the edges of their continental shelves rather than their present coastlines) looked, as Wegener put it, like a sheet of newspaper torn irregularly in two: as if they had moved apart and the Atlantic had opened up between them. This was just what Wegener believed had happened. And he claimed that there was evidence that this horizontal movement of the Earth's crust was still continuing, just like the uncontroversial vertical movement in Scandinavia: longitude measurements made on successive expeditions to Greenland (he himself had been on one) suggested that it was very slowly moving away from Europe. Continental displacement therefore seemed to be endorsed by the universally respected principle of actualism; it was, in Lyell's famous phrase, a "cause now in operation" (the measurements were later judged too imprecise to be decisive, but eventually the movement was confirmed by GPS, though at a rate even slower than Wegener imagined).

Being a meteorologist as well as a geologist, Wegener was equally impressed by the evidence for the different sequences of climatic change that continents had undergone in the course of their long-term histories, as inferred from the stratigraphical and fossil record of each: for example, during the Carboniferous period Europe and North America were apparently tropical but India was glacial, whereas by the Pleistocene these climates had been reversed. Such huge climatic changes could no longer be attributed to a uniform cooling of the whole globe, but they made good sense if each landmass had moved more or less independently through different latitudes in the course of its history. Wegener acknowledged the need to find an adequate physical cause for all this dramatic continental displacement, but he relegated his discussion of this to a final chapter, because he insisted that establishing its historical reality should be given priority. Alluding to the earlier example of gravitation, he conceded in the last edition of his work (1929) that "the Newton of displacement theory has not yet appeared," and he did not cast himself in that role (the following year,

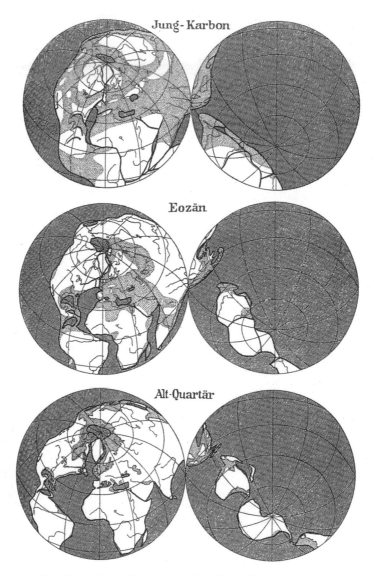

**FIG. 10.2** Alfred Wegener's provisional reconstruction of the Earth's continents at three successive periods in its history: late in the Paleozoic era (*Jung-Karbon*, late Carboniferous), early in the Cenozoic era (*Eozän*, Eocene) and around the start of the Pleistocene Ice Age in the geologically recent past (*Alt-Quartär*, early Quaternary). The shaded parts of the continents represented shallow seas (such as the "continental shelf" that now surrounds the British Isles). These maps, published by Wegener in 1922, showed what he claimed as the splitting up of an early super-continent, with the gradual opening of the Atlantic Ocean. Wegener noted that whereas "palaeo-climates" (based on the evidence of rocks and fossils) could suggest approximate "palaeo-latitudes" for the continents, "palaeo-longitudes" were unavoidably conjectural. The margins of the *present* continents and their major rivers were shown here just for reference.

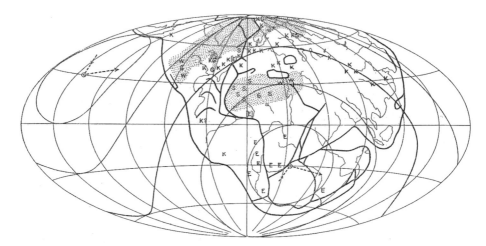

**FIG. 10.3** "Ice, swamps and deserts at the time of the Coal": a tentative reconstruction of world climates in the Carboniferous period, published by Wegener and his father-in-law Wladimir Köppen in their *Climates of the Geological Past* (*Die Klimate der geologischen Vorzeit*, 1924), a book that supplemented Wegener's better-known work on continental displacement or "drift." The distribution of traces of glacial ice (*E*) in inferred polar regions, and of coal (*K*) in inferred tropics, and between them arid areas (stippled) with salt (*S*) and gypsum (*G*) deposits and desert sandstone (*W*), all made much better sense if the present continents were at that time united, as shown, in a single vast super-continent (*Pangaea*), with the south pole close to the east coast of southern Africa, and the north pole out in the Pacific, far to the west of North America. (This map superimposed the inferred Carboniferous latitudes on a grid based on the *present* positions of the landmasses that border the present Atlantic.)

tragically, he died on Greenland's inland ice sheet while on another expedition).

## CONTROVERSY OVER CONTINENTAL "DRIFT"

Wegener's displacement theory was heady stuff, and from the start it divided Earth scientists into those prepared at least to give it a hearing, and those who dismissed it outright. The case for mobilism was not helped when in the Anglophone world the theory became generally known as "*continental drift*," which suggested the inappropriate analogy of drifting ice-floes or icebergs. The disagreement sometimes seemed to be between those familiar with the geological and biological evidence, and those who best understood the Earth's large-scale physics: between geologists and biologists on the one hand, and physicists and geophysicists on the other. So, for example, the British physicist

Harold Jefferies, in his formidably mathematical book on *The Earth* (1924), dismissed Wegener's theory as physically impossible; but many British geologists, "once bitten" by Kelvin, were "twice shy" about any such dogmatism from any physicist. In the United States, however, Wegener's fiercest critics included not only physicists but also many geologists. For example, the leading stratigrapher and palaeontologist Charles Schuchert of Yale insisted that former land-bridges were quite adequate to account for the global distribution of plants and animals, both fossil and living. As he later told Holmes, who in the meantime had become a strong mobilist, "Gondwana is a fact, but I still have to get rid of it, or else go into the Wegener following, and heaven knows I will not trail into that trap!"

The sharpest contrast on this issue was not in fact between geologists and geophysicists but between the United States and the rest of the scientific world. At an important meeting in England in 1922 the opinions of a variety of scientists ranged from a respectful scepticism to a cautious agreement that the theory was well worth further investigation. In contrast, at a similar meeting in New York in 1926, all the scientists were virulently hostile—Schuchert was one—apart from Willem van Waterschoot van der Gracht, the Dutch-born oil geologist who had organized it and who had realized the theory's enormous implications. Several factors, the relative importance of which it is difficult to assess, contributed to the almost universal opposition by Earth scientists in the United States. Prominent among them was a widespread antipathy to Wegener's eloquent advocacy of his own theory—just the grounds on which Lyell had been criticized a century earlier!—and a belief that he ought to have evaluated the alternative explanations more soberly and even-handedly. Another factor was a refusal to accept displacement unless and until an adequate physical cause for it had been found, though this ignored Wegener's insistence that judging its historical reality should have priority. Yet another was that Wegener, like Suess and other European geologists, cited the published reports of many other scientists from around the world, rather than emphasizing his own first-hand fieldwork. All these factors reflected a distinct contrast between Americans and Europeans in their scientific style and their concept of sound scientific methods. Furthermore, the Americans' geological experience was often limited to their own vast continent, whereas Europeans' knowledge often covered continents beyond their own, thanks to their contacts and colleagues in present or former colonies around the world. Finally, two other factors should not be ignored or swept under the carpet. American scientists at this time were justifiably proud of their recent rise to prominence in the scientific world and were therefore reluctant to concede that any European had beaten them in proposing a new theory of such potential importance. And Wegener was not just a European, but a German, at a time when Germans were being deliberately cold-shouldered or even excluded from international collaboration in some of the sciences, in vengeful reparation for the horrors of the recent war.

In fact, one apparent exception to this almost universal American hostility to mobilism confirms the general rule. The distinguished Harvard geologist Reginald Daly was an early mobilist in America, but he was a Canadian by birth and education, well traveled and fluent in French and German (he read Wegener's book in its original German). While working in South Africa in 1922 he met the leading South African geologist Alex Du Toit, who was already a strong advocate of mobilism. Under his guidance Daly saw some of the South African field evidence, such as the ancient tillites that suggested that the climate there had been glacial in the Carboniferous period. He became convinced that only the fragmentation of Gondwanaland and the dispersal of its parts was adequate to account for this. Back in America he proposed that Du Toit, as an internationally respected expert on African geology—one side of Wegener's prime example of the putative displacement of continents—should be given the resources necessary to compare it with the geology of South America on the other side. However, his proposal for Carnegie funding for Du Toit was turned down until he resubmitted it, presenting it no longer as a test of Wegener's theory but as an evaluation of both displacement and land-bridges, as alternatives of equal plausibility. But Du Toit in effect ignored this injunction when in 1923 he carried out extensive fieldwork in South America (making good use, of course, of the many geologists working there). His report, published later in the United States, presented overwhelming evidence that there was not only a close fit between the edges of the continental shelves that faced each other across the South Atlantic, but also a close match between the geological *histories* of the two continents, all the way from the Precambrian through the end of the Cretaceous period, after which they might have begun to split

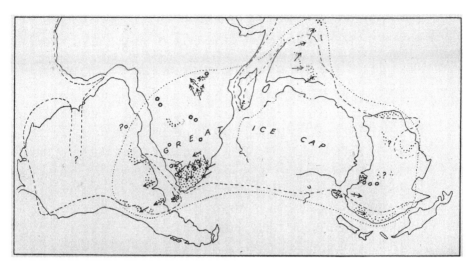

**FIG. 10.4** Alex Du Toit's tentative reconstruction of Gondwanaland in the Carboniferous period, as published in *Our Wandering Continents* (1937). A "Great Ice Cap" (outlined with a dotted line)—comparable in area to the far more recent Pleistocene ice sheets in the Northern Hemisphere—was reconstructed from the evidence of tillites (small circles), other glacial deposits (stippled) and scratched bedrock showing the direction of ice movement (arrows). (The dashed line indicated the approximate coastline of the super-continent itself, beyond which deposits of this age were marine.) The distribution of these traces of glacial conditions was far more difficult to understand if at the time these continents had been in their present positions on the globe. The position of India was particularly striking, with its clear signs of ice movement northwards or *away from* the present equator. Like Wegener before him, Du Toit fitted the continents together along the edges of their continental shelves (as far as these were known), with their present coastlines shown just for reference.

apart. Daly used all this evidence to promote the mobilist cause in *Our Mobile Earth* (1926), but it had little effect on geological opinion in the United States. Du Toit's later book on *Our Wandering Continents* (1937) fared little better there, but it had a major impact on geologists elsewhere. It summarized the rapidly improving case for mobilism in a way that was particularly persuasive in the eyes of geologists who, like Du Toit himself, were working in the Gondwanaland continents or who were at least familiar with their geology.

In Europe too the theory gained growing support, not least after Holmes—by now a highly respected scientist who had initially been a sceptic—became convinced that mobilism was fundamentally sound. In 1928 he suggested a physical mechanism for the displacement of continents that, although not wholly original, was more plausible

than earlier conjectures: he concluded that "its general geological success seems to justify its tentative adoption as a working hypothesis
of unusual promise." It incorporated the latest thinking about the
Earth's heat budget in the light of what had become known about its
radioactivity. Holmes suggested that in the deep mantle underlying
the Earth's crust there might be a system of huge convection currents. These might be capable—on the immensely lengthy scale of the
Earth's history—of slowly pulling apart continents such as Africa and
South America and creating new oceans such as the Atlantic between
them. On this model the continents were not, implausibly, *drifting* like
icebergs through seas of solid rock, but being slowly *dragged along*
by underlying materials that in the long run were viscous. Holmes's
theory certainly made mobilism much more believable in the eyes of
many Earth scientists in Britain and other European countries. He
summarized it towards the end of the Second World War in the final chapter of his *Principles of Physical Geology* (1944). This highly
successful book, written primarily for students of the Earth sciences,
enticed many of them (I was one) in the postwar period to ignore the

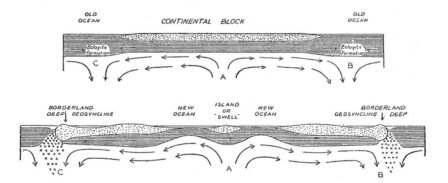

**FIG. 10.5** Arthur Holmes's conjectural theory of continental displacement, showing how huge
convection currents in the deep "mantle" rocks below the Earth's crust might be capable of
dragging apart a single "continental block," to form two new continents separated by a new
ocean (and a mid-ocean "swell" marked by islands). This idealised diagram was published in
1931 in an early and explicitly provisional version of his theory; he put it forward as an idea that
could be improved, as it duly was—though without adequate acknowledgment—in the theory
of plate tectonics developed in the 1960s. Holmes's theory, even in this early form, showed
that continental displacement should not have been dismissed on the grounds that there was
no conceivable causal mechanism for it: continents such as Africa and South America might
have been forcibly dragged apart in the course of the Earth's history, rather than—implausibly—
"drifting" through a resistant solid substrate.

scepticism of most of their elders and betters, and to conclude that mobilism was likely to be on the right lines.

Holmes's convection model was supported early on by some of the first research to move the focus of global geophysics from the continents to the oceans. For example, the Dutch geophysicist Felix Vening Meinesz had devised an instrument that could measure the force of gravity with unprecedented precision, and in 1923 he took it on a Dutch submarine down to depths where it was not disturbed by wave motion. He mapped the gravitational field along the deep oceanic trench close to the island of Java and interpreted the tiny anomalies as evidence for a local downwarping of the Earth's crust. In 1934, in the light of Holmes's theory, he reconfigured this as evidence for an underlying downward convection current at the junction between two segments of the Earth's crust, associated with the arc of active volcanos and frequent earthquakes along the length of the Dutch East Indies (now Indonesia). This was soon confirmed by further research of the same kind elsewhere, by American geophysicists among others, which supported and steadily improved the idea of subcrustal convection. So the continuing opposition of most Earth scientists in America to the very idea of continental displacement was not due to the lack of a plausible physical cause to account for it.

Instead, much of the argument among Americans continued to focus on the geological evidence for a former Gondwanaland, and for a former northern super-continent ("*Laurasia*") that on the mobilist interpretation might once have united North America and Eurasia. Schuchert was among those who, in contrast, continued to argue that the biological evidence could be adequately explained in terms of "*isthmian links*": former land-bridges perhaps transient and slender enough to be somehow compatible with the demands of isostasy. After Schuchert's death in 1942, George Gaylord Simpson, an expert on fossil mammals (and a leading contributor to the "modern synthesis" or neo-Darwinian evolutionary theory beginning to take shape at this time), reinforced the same line of reasoning with the claim that the present isthmus at Panama was just such a land-bridge: its emergence in the geologically recent past had enabled North American mammals to invade South America and wipe out many of the indigenous species. And Simpson argued that, if earlier land-bridges were to be disallowed on geophysical grounds, then the occasional natural rafting of terrestrial organisms across oceans would still be adequate to explain their distribution. One of Schuchert's former students had summarized American opinion by noting that "Gondwana is on the wane alright," and with it the necessity to invoke any continental mobility at all. Du Toit was understandably exasperated by what he saw as the double standards being applied by many American scientists, who demanded more rigorous evidence for mobilism than what they offered for fixism. As another South African geologist commented to him in 1940—just when the United States was remaining resolutely neutral in the increasingly global war against Nazi Germany and its allies—"The Americans are about the toughest isolationists in existence, geologically as well as politically."

## A NEW GLOBAL TECTONICS

The Second World War, more than the First, directed much scientific research towards military objectives. In the long run this benefited the Earth sciences, in that American geophysicists were diverted into tackling naval problems—such as the detection of enemy submarines—that demanded better knowledge of the oceans. But as the end of the hot war merged seamlessly into the start of the Cold War, most of this new knowledge remained secret and, at first, unavailable to the burgeoning science of oceanography. Meanwhile, although Britain was greatly weakened by the war, it saw a new line of non-military research that created a new tool for reconstructing the Earth's history. Like radiometric dating, this was an almost inadvertent product of research in physics. The Cambridge geophysicist Edward ("Teddy") Bullard suggested that the Earth's magnetic field might be related to the hypothetical convection currents deep in its interior. If so, its changes over time might be recorded in changes in the magnetism preserved within igneous rocks. As intensely hot liquid magma such as volcanic lava cooled to form such rocks, some of the minerals that crystallized out (particularly the iron mineral *magnetite*) would in effect pick up the surrounding magnetic field. The Earth's magnetic field at that time and place would therefore be frozen in this "*palaeo-magnetism*." Newly sensitive instruments could detect this faint "fossil" trace, which in turn could indicate the approximate latitude at which the rock was formed (the magnetic poles, although changing continuously in position, remain quite close to the geographical poles and were assumed to have done so in the past). In 1954 some of Bullard's associates at Cambridge published the results of their palaeo-magnetic measurements of the palaeo-latitudes of British rocks of different geological ages, from the Precambrian towards the present. This suggested that the bit of continental crust that now forms Britain had changed gradually but radically in its latitude during the past few hundred million years, confirming what the geological evidence of its changing climates already implied.

For a time it seemed that these changes in latitude could be explained by "*polar wandering*," by which the planet's axis of rotation might have changed relative to all the continents and oceans at the Earth's surface, without any change in their positions relative to each other. But Ted Irving, a Cambridge geophysicist who had moved to Australia, plotted the palaeo-latitudes of similar sequences of rocks from other continents, including those of Gondwanaland. It then seemed as if the polar positions they indicated gradually diverged as they were traced back in time, which implied that, conversely, the continents must have moved apart on the Earth's surface in the course of its history. This was yet another strong piece of evidence in favor of what Anglophone Earth scientists now referred to as "continental drift" (despite that term's misleading overtones). And since it was evidence based on physics, yielding quantitative results derived from delicate instrumentation, it appealed to geophysicists much more than the necessarily qualitative evidence of palaeo-climates that geologists had been invoking ever since Wegener. It was later reinforced when Bullard and his Cambridge colleagues combined the latest oceanographical surveys of the continental shelves with the capabilities of early computers. They convinced all but the most sceptical among their fellow geophysicists that the "fit" between Africa and South America, which—again, ever since Wegener—had been claimed as primary evidence for mobilism, was much too precise to be dismissed as coincidental. This and other new research at last began to swing the balance of opinion among Earth scientists, even Americans, in favor of mobilism. And those in the Soviet Union, who had been almost as unanimously hostile to mobilism as the Americans, and for somewhat similar reasons, also began to consider it more favorably.

During the 1960s "continental drift" came to be accepted, by all

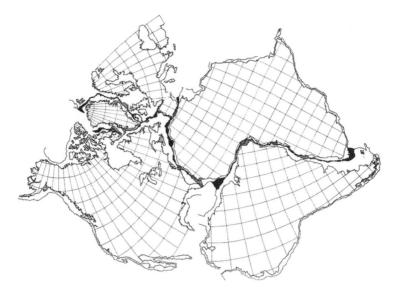

**FIG. 10.6** The close fit between the continents now bordering the Atlantic, on a mobilist interpretation of their history, as published in 1965 by Edward Bullard, Jim Everett, and Alan Smith. Their Cambridge colleague the physicist Harold Jeffreys, who was a forceful critic of any such theorizing, had denied that there was as close a fit as mobilists claimed. This was the first time that computers were used to test the claim, taking accurate account of the spherical geometry of the Earth's surface and the latest oceanographic data on the submarine topography of the continental margins. The fit was best at the present 500 m contour on the slopes of the continental shelves: the irregular black strips (colored on the original) indicated the areas where there was either a slight gap or small areas of overlap; the latter were explicable as the result of deposition (for example in the Niger delta) during the time since the *"tectonic plates"* underlying the continents started to move apart. The Iberian peninsula (Spain and Portugal) was shown rotated from its present orientation, but there was independent geological evidence for this; Central America and the Caribbean remained areas of uncertainty. The present shorelines of the continents, and their grids of present latitude and longitude, were shown just for reference.

but a few diehards, as a real major feature in the Earth's long-term *history*, not least because during the same few years most Earth scientists became convinced that a physically adequate *cause* had been identified. Holmes's theory of convection currents in the Earth's deep interior was reformulated and improved—though often with inadequate acknowledgment—by incorporating the massive new information about the oceans that was derived from the postwar compact between American oceanographers and geophysicists and the United States navy. For example, as one such incidental product of the Cold War, huge *"mid-ocean ridges"* with submarine volcanoes rising from the deep ocean floor were mapped in detail, notably down the whole length of the Atlantic (rising above sea level here and there, as volcanic islands such as Iceland, the Azores, and Ascension). Along the crest of each of these ridges a lengthy "rift valley" was discovered, comparable to the one seen on land in East Africa; and this was identified as the line along which new magma was currently being extruded as volcanic lava. The study of palaeo-magnetism in piles of successive lava flows had already revealed, very surprisingly, that the Earth's magnetic field must have *reversed its polarity* ("north" becoming "south" and "south" "north") quite frequently, by geological standards, in the course of time. This provided a new kind of relative dating, closely analogous to the stratigraphical, for charting the more recent periods of the Earth's history; and when combined with radiometric dating for the same rocks this distinctive sequence of magnetic reversals could even be given approximate dates.

A crucial breakthrough came with the discovery that the ocean floor on either side of a mid-ocean ridge, when mapped in detail, was marked by symmetrical tracts magnetized with the same alternating sequence of polarities. The Cambridge geophysicists Fred Vine and Drummond Matthews interpreted this as evidence that new ocean floor material was being formed continuously by the extrusion of new lavas at mid-ocean ridges. This in turn was taken as evidence that a convection current was rising along that line and slowly dragging the underlying rocks apart in both directions. Extended over geological time, this process could, for example, have dragged Africa and Europe apart from the Americas, opening the Atlantic between them. This idea was then combined with an interpretation of oceanic trenches and island arcs—developing the earlier ideas of Vening Meinesz and others—as the complementary lines along which the rocks underlying the ocean floor were being "*subducted*" or dragged downwards by descending convection currents, back into the deep interior. It was a literally cyclic and potentially steady-state system, which would have delighted Lyell.

The upshot of all this research, combining the "fieldwork" of underwater oceanographic investigation with highly theoretical geophysics, was a modified version of Holmes's causal explanation of continental displacement. What were now treated as "drifting"—the term was less appropriate than ever—were no longer the continents as such, but

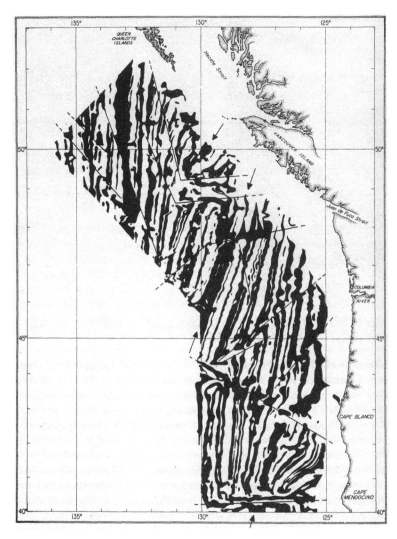

**FIG. 10.7** A map, published in 1961, of a part of the eastern Pacific off the coast of British Columbia and the states of Washington and Oregon, showing the stripes of ocean floor marked by
magnetic anomalies alternately positive (black) and negative (white), as charted by instruments
on board an oceanographic research vessel. This *"zebra pattern"* was later interpreted as a record
of the history of the repeated reversals of the Earth's magnetic field, preserved by being "fossilized" in the rocks that had been continuously extruded as liquid lava on the ocean floor; the
stripes were thus a *historical* record analogous to the sequence of strata in a pile of rock formations. The pattern of stripes was symmetrical on either side of the Juan de Fuca Ridge (marked
by the broad black stripe in the center of the map, with two arrows pointing to it): this was
identified as the line along which, in this region, the new magma was currently being extruded
to form new ocean floor. In the 1960s this interpretation of the magnetic anomalies was treated
as evidence for the geophysicists' theory of *"plate tectonics"*; it provided a satisfactory causal
explanation of the slow lateral movement of continents, for which geologists had already found
strong evidence in the Earth's long-term history.

wider *"tectonic plates"* or segments of the Earth's crust that might or might not carry continents, as it were, on their backs. So, for example, it was claimed that one of the plates that was being added to at the mid-Atlantic ridge extended westwards across the South Atlantic and South America; at the western edge of that continent it was said to be piling up to form the Andes, because it was overriding an adjacent plate underlying the eastern Pacific, which was being subducted along the same line. By the later 1960s there was enough geophysical evidence for tectonic plates such as these to be mapped, at least provisionally, around the whole globe, together with the inferred rates at which new material was currently being formed or older material subducted along each plate boundary. This new version of Holmes's theory came to be known as *"plate tectonics."* It was treated as super-

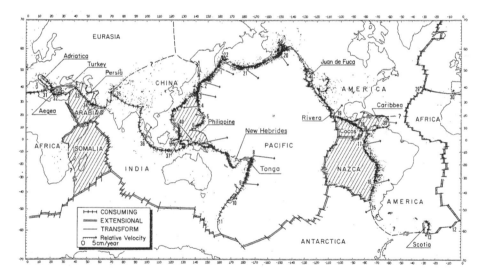

FIG. 10.8 A map of the world (on Mercator projection) published in 1973 by Xavier Le Pichon and his colleagues, showing the six great *tectonic plates* that he had proposed in 1968 to explain the then known evidence, together with smaller plates (shaded) defined subsequently. The map marks the "extensional" boundaries where the plates were thought to be pulling apart, and the subduction or "consuming" boundaries where they were being crushed together; there was a dense concentration of recorded earthquakes (their epicenters marked by fine dots) on and near the latter. It was even possible at some points to estimate the current rates of relative movement of the plates (arrows of measured length). This map also shows how the theory of plate tectonics transcended that of continental displacement or "drift," by shifting the focus of attention from the continents themselves to the much larger "plates" that might or might not carry continents as it were on their backs.

seding or replacing earlier ideas of continental displacement, though in fact the continuities were far greater than the forceful advocates of the new theory were generally prepared to concede. Plate tectonics was primarily the creation of geophysicists and oceanographers, who were often openly scornful about the less quantitative and mainly land-based work of the geologists and biologists who had provided most of the evidence for the *historical reality* of continental displacement. Conversely, however, plate tectonics did provide a fuller and more satisfactory account of the *causal processes* underlying it, the lack of which had for so long been treated as a major objection to mobilism of any kind.

The theory of plate tectonics was constructed by the collaborative work of many scientists (far too many to have been named adequately in this necessarily sketchy summary). Among them Americans were the most prominent, but they had to overcome the greatest barriers of scepticism about any kind of mobilism. It is significant that when in 1960 Harry Hess of Princeton, one of the most important figures in this story, proposed the provisional model of plate tectonics that inspired many others in the following years, he felt it necessary or prudent to call it "*geopoetry*," as if to excuse an otherwise unacceptable piece of speculative theorizing. He contrasted it with Wegener's less satisfactory theory, but overlooked or ignored Holmes and Du Toit and all the other scientists who had worked hard to improve on Wegener during the intervening decades. In the course of the 1960s, Earth scientists in the United States performed a startling collective volte-face on mobilism and rapidly became some of the most ardent advocates of plate tectonics, even to the point of excoriating the few remaining fixists as fiercely as they or their predecessors had criticized the early mobilists. Plate tectonics was trumpeted as a major "scientific revolution"—Thomas Kuhn's recent *Structure of Scientific Revolutions* (1962) was being widely discussed among scientists as well as by historians and philosophers of science—as if the possibility of mobilism had only just been broached and as if it had suddenly and completely transformed the Earth sciences. This left European geologists, many of whom had long been inclined to favor mobilism in some form, somewhat bemused. (As an Earth scientist, but not one directly involved, I watched all this from the sidelines during the 1960s).

While perhaps resenting all this triumphalism, Earth scientists working in the Gondwanaland continents justifiably felt that at long last they had been vindicated. And among geologists and palaeontologists everywhere, at least those of younger generations, there was a palpable sense of relief. They felt that they were at last permitted to take the massive changeability of global geography into account in their reconstructions of the Earth's history, without being sniped at by

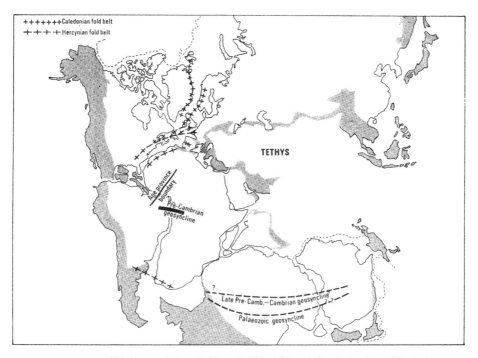

**FIG. 10.9** Global geography in early Mesozoic (Triassic) time, before the break-up of the single super-continent Pangaea. This reconstruction, published in 1973 by the English geologist Anthony Hallam, was based on one of the series of such maps compiled by Alan Smith and Joe Briden for successive periods in the Earth's history, incorporating what was then very recent research on plate tectonics; but its general character is quite similar to Wegener's earlier reconstructions (as usual the continents are all shown with their present coastlines, just for reference). The older mountain ranges or "fold-belts" that had long been named *Caledonian* and *Hercynian* were much more plausible in these positions than when, in late 19th-century reconstructions, they had been shown extending right across the present Atlantic. The vast former ocean named *Tethys* was taken to have been eliminated later in the Earth's history, when for example Africa (with Arabia) and India both moved northwards and collided with Asia. The stippled areas on this map are those affected by subsequent (Cenozoic) mountain-building movements, and their form in Triassic time was therefore highly uncertain. By the 1970s geological exploration of Antarctica was beginning to give better evidence for incorporating that vast continent into reconstructions such as this.

geophysicists for invoking what was said to be physically impossible. During the rest of the 20th century the reconstruction of past global geographies became almost a matter of routine, as palaeo-magnetic indications of former latitudes were reported from an ever wider range of localities around the world and from rocks of ever more diverse geological ages. It was found that palaeo-latitudes could be derived not only from igneous rocks such as ancient lava flows but also from some sedimentary rocks: as grains of magnetite had settled on the sea floor along with ordinary grains of sand (quartz), they had oriented themselves like tiny compasses in the Earth's magnetic field as it was at that time and place. Measuring the palaeo-latitudes of rocks became almost as straightforward as finding their radiometric dates. Former longitudes were more of a problem, as they had been for Wegener; but other kinds of evidence, derived from the geology of both continents and oceans, made it possible to construct increasingly plausible global maps for different geological periods. These in turn made much better sense of the distribution of fossil organisms, both terrestrial and marine, in past floral and faunal "provinces" like those of living plants and animals. And palaeo-climates were reconstructed with increasing confidence, not least after the discovery that the former temperatures of the oceans could often be inferred from the isotopes of oxygen incorporated in the shells of fossil organisms.

The effect of all this research was to reinforce still further the sense that the Earth's history had been highly contingent throughout, and therefore utterly unpredictable even in retrospect. Continents had apparently converged and coalesced to form super-continents such as Gondwanaland or even at times a single huge "*Pangaea*" ("whole-Earth"), and then split apart again. New oceans such as the Atlantic had opened up, then closed, and then reopened in a different position, so that fragments of one former continent were left attached to another (there was evidence, for example, that the northwestern fringe of Britain belonged, geologically, to an earlier North America). Earlier oceans such as the great seaway called "*Tethys*" between Africa and Eurasia had completely closed or just left the Mediterranean as a small remnant. As tectonic plates were crushed together, a succession of great mountain ranges such as the Alps, the Andes, and the Himalaya had arisen; older ranges had been reduced by long-continued erosion to mere hilly stumps such as the Scottish Highlands and the Appalachians. By the late 20th century, even more than before, scientists had good reason to regard the Earth as a highly dynamic system, and its history as not only unimaginably lengthy but also amazingly eventful. The next and last chapter of this narrative explores how, also during the 20th century, this dynamic and eventful global history came to be integrated into an even wider picture, as the Earth was increasingly treated as a planet in a cosmic context.

# 11

## One Planet Among Many

### EXPLOITING THE EARTH'S CHRONOLOGY

During the 20th century and into the early 21st, the "geochronology" that was made possible by radiometric dating allowed the dramatic changes indicated by continental drift and plate tectonics to be reconstructed on an "absolute" timescale of many millions—and tens and hundreds of millions—of years. But radiometric dating also had a major impact on geologists' understanding of many other aspects of the Earth's history, and on every part of it from the relatively recent to the most ancient.

As an example from the more recent end, the dating of early human fossils and human artifacts tied geology into archaeology even more closely than in the earlier study of prehistory. Geologists and archaeologists collaborated in the reconstruction of an approximately *dated* history of humanity, all the way from the emergence of the human species from its apparent primate ancestors in the late Cenozoic era, to the development of literate civilizations from their pastoral and hunter-gathering forerunners only a few millennia ago. Human origins could be plotted with increasing confidence, all the way from the somewhat human-like fossil *Australopithecus* ("southern ape") that was first found in southern Africa in the 1920s, to the extant species *Homo sapiens*. In place of the singular "missing link" that had been anticipated earlier, the gap between them was filled by a profusion of near-human forms that were found in increasing variety, particularly but not only in Africa, from the 1950s onwards. Just one among them was the *Homo erectus* (formerly known as *Pithecanthropus* or Java Man) that was widely regarded as our likely ancestor, in a way that Neanderthal Man and others were not. And radiometric dating of human fossils from around the world made it possible to trace not only the evolution of the human species itself, but also its apparent dispersal out of Africa across Europe and Asia as far as Australia, and eventually to the Americas and at the last moment to the scattered islands of the Pacific. This lengthy human prehistory could now, at least in principle, be correlated with the story of successive glacial episodes and milder interglacial conditions on the northern continents during the Pleistocene period, and the corresponding climatic fluctuations at lower latitudes and on the southern continents. It became possible to reconstruct the changing local environments in which early humans and their forerunners had lived and evolved. The details of all this complex history were often highly controversial. But the broader outlines became progressively clearer in the later 20th century and into the 21st, thanks to the cumulative evidence of new fossil specimens recovered from ever more localities around the world, combined with more precise measurements of their geological ages and more confident reconstructions of the physical activities and mental capacities of the creatures they represented.

The impact of radiometric dating on the reconstruction of the middling part of the Earth's history, from the Cambrian period to the Pleistocene—which geologists in the 19th century had explored so successfully—was equally far-reaching, although not as sensational or newsworthy outside the circles of the relevant scientists. They were able to allocate the "millions" in their bank of deep time, with increasing precision, to the successive geological periods and to their component epochs (a term that was now applied to spans of time rather than decisive moments). The boundaries between these named spans of time—say, between the Silurian period and the Devonian—were recognized as being essentially conventional and matters of convenience; during the 20th century they came to be fixed by formal international

**FIG. 11.1** The fossil footprints of human-like individuals, found in 1978 by the Kenyan scientist Mary Leakey. They were preserved on a surface of volcanic ash at Laetoli in Tanzania and dated at about 3.6 Ma in the Pliocene epoch or late Cenozoic era. The report of the find, in which this photograph was published in 1979, stated that "it is immediately apparent that the Pliocene hominids at Laetoli had achieved a fully upright, bipedal and free-striding gait," although their brains (as inferred from fossil skull bones found at sites of about the same age) were probably comparable to those of chimpanzees. This implied that a bipedal habit—which would have freed the forelimbs for other tasks—developed long before the dramatic increase in brain size that was thought to characterize the line leading to modern humans. This was important new evidence on human origins, and all the more valuable for its fairly accurate radiometric dating.

agreement among scientists, and no longer just by the decision of some powerful Murchison-like individual. Yet geologists recognized that some of these boundaries might also be natural, if in fact they marked times of rapid or radical change in the organisms preserved as fossils. An increasingly accurate timescale made it feasible for the first time to plot the points in deep time at which new groups of animals and plants appeared in the fossil record or disappeared from it, and then to estimate the rates of change caused by the evolution of new forms of life and the extinction of older ones. Only the calibration made possible by radiometric dating could hope to show, for example, whether the major discontinuities in the fossil record at the ends of the Palaeozoic and Mesozoic eras—which had already been apparent to Phillips and his contemporaries in the mid-19th century—were due to episodes of relatively sudden mass extinction, or, as Lyell had thought, simply to major gaps in the preserved record of what in reality might have been a steady rate of very slow and gradual change.

One early sign of this trend towards calibrating the history of the Earth and documenting the history of its life—even before rapidly improving digital technologies made it easier to process the huge bodies of data that were involved—was the publication of two important collaborative compilations, *The Phanerozoic Timescale* (1964; the technical term will be explained shortly) and *The Fossil Record* (1967). The driving force behind both was Brian Harland, a Cambridge geologist with an exceptionally clear vision of the outstanding major problems in his science (as one of his palaeontological colleagues I had a small part in the second of these projects). Taken together, they provided data that seemed to support what some scientists had long suspected: that the history of life on Earth might not have been as smooth and uniform as Lyell and Darwin had supposed. The catastrophists among their 19th-century contemporaries might have had a more realistic picture of its highly eventful course. A revival of catastrophism might be overdue.

## THE RETURN OF CATASTROPHES

This, however, was not an inference that was generally welcomed among Earth scientists during much of the 20th century. The leading German palaeontologist Otto Schindewolf, who was an exception,

coined the term "*neo-catastrophism*" in 1963, but the idea behind it
was generally rejected; it did not help his cause that he was well known
for his non-Darwinian version of evolutionary theory. Particularly
among Anglophone scientists, there was still a strong inclination to
dismiss any suggestion that there might ever have been "catastrophic"
events of an intensity beyond any observed at the present day or
recorded in human history. Any suggested rehabilitation of catastrophism was forcefully rejected as contrary to good scientific method,
as a deplorable throwback to less scientific times, or even—in gross
ignorance of its history—as a sign of covert biblical literalism. Lyell's
eloquent rhetoric in favor of "absolute uniformity" retained its persuasive power more than a century after it was first propounded.

A conspicuous but far from unique example of this prejudice—
for such it was—concerned the geology of the Spokane River basin
in a remote part of Washington State. In 1923 the geologist Harlen
Bretz described what he called "*channeled scabland*" (a ranchers' term
for barren ground) with deeply eroded dry valleys and large erratic
blocks, yet no signs that the area had ever been glaciated. He argued
that only a sudden, violent, and extremely massive flood, some time
in the Pleistocene period, could account for these strange features. But
at a meeting in 1927 other American geologists forcefully dismissed
Bretz's "Spokane Flood," ostensibly because he suggested no cause for
the event, but more fundamentally because his claim was thought to
violate the canons of "uniformity"; this was, significantly, only a year
after another gathering of American geologists had rejected Wegener's
ideas about mobile continents. Even worse, Bretz's flood was thought
to resemble the early kind of diluvial theory that had equated a supposed geological Deluge with the biblical Flood; and this at a time
when those who called themselves "fundamentalists" were beginning
to flex their political muscles in the United States, to the understandable alarm of scientists there. So it is not surprising that Bretz was
marginalized and his ideas dismissed. But in 1940 another geologist,
Joe Pardee, while studying the region upstream from the channeled
scabland, found abundant evidence that a vast former lake had indeed been suddenly drained: the melting of a natural dam of glacial
ice had released a huge volume of water all at once through a narrow
gap into the Spokane River basin (if geologists had been more familiar
with their own history, they might have recalled the famous case of the

natural dam burst in the Val de Bagnes in the early 19th century, as a comparable event on a smaller scale). Not until 1965 was Bretz fully vindicated, when an international group of geologists studied the channeled scabland at first hand, were duly convinced, and sent him a congratulatory telegram saying "We are now all catastrophists."

The opposition among geologists to any kind of catastrophism did indeed die down, though more slowly than the telegram suggested. In the course of the 1970s the possibility that the deeper reaches of the Earth's history, long before the Pleistocene, had been marked by far greater catastrophes than the Spokane Flood began to be taken seriously by Earth scientists, even in the United States. In 1982, for example, Kenneth Hsü, a prominent geologist born in China, educated in the US but working in Switzerland, argued—without damaging his scientific reputation—that a lot of otherwise puzzling evidence around and beneath the Mediterranean Sea could be explained by inferring that during the later Cenozoic era this vast basin had been cut off from the world's oceans, had gradually dried out (as is now happening, on a far smaller scale, to the Dead Sea), and had then been rapidly flooded when the present Strait of Gibraltar was breached. It was of course an impeccably natural cause for a huge and catastrophic event. In the 1990s a much more recent sudden flooding of the Black Sea basin, by the breaching of the narrow Bosphorus at Istanbul, was suggested as a possible historical source for the flood stories preserved in the literate records of the ancient Near East. In this case the scientists were castigated for potentially giving comfort to biblical literalists, by daring to suggest that Noah's Flood might have had some historical basis (if these critics had known the history of their own science they would have recalled the highly respectable precedents for this kind of "euhemerist" inference, for example in Suess's great geological synthesis in the late 19th century).

The possibility that there had been catastrophic natural events in the Earth's deep past should have been of great interest to evolutionary biologists, but for much of the 20th century they were almost wholly absorbed with updating Darwin's ideas about how new *species* originate: the celebrated decoding of DNA in 1953 was a landmark in their consistent (and highly fruitful) pursuit of causal questions about "micro-evolution." They tended to dismiss fossil evidence about any larger features of the evolutionary story as almost useless, and many of them openly treated the science of palaeontology as inferior and irrelevant. In the 1970s, however, a group of younger American palaeontologists began to use the rapidly increasing potential of computers to analyze the kind of detailed information about the fossil record that has already been mentioned. In 1975 they founded the new periodical *Paleobiology* (a title that in fact had been anticipated by Schindewolf in the 1920s), to promote research that would be directed—more closely than in conventional palaeontology—towards solving the *biological* "big issues" raised by the fossil record. They hoped to bring palaeontology back to the "high table" of debate about evolutionary theory, from which they felt it had been unfairly and unwisely excluded. In 1982 two of them, Dave Raup and Jack Sepkoski, summarized their exhaustive statistical analysis of the known fossil record, and identified five points at which there was strong evidence for episodes of mass extinction. Of these five, the two events with the greatest apparent rates of extinction were at the boundaries between the Permian period and the Triassic, and between the Cretaceous and the Tertiary. This confirmed, with a hugely enlarged body of evidence, what Phillips had judged over a century earlier, when he used these two points to define the boundaries between his three great eras in the history of life, Palaeozoic, Mesozoic, and Cenozoic.

According to this analysis, the greatest of all these mass extinction events had been at the Permian-Triassic boundary; that is, between the Palaeozoic and Mesozoic eras. This was no surprise to palaeontologists familiar with this part of the fossil record. The boundary had long been a matter of discussion among geologists. Only a careful comparison of the relevant rock formations in different regions could show whether the extinction of Permian forms of life and the later appearance of Triassic ones had been sudden or gradual. In 1937 an International Geological Congress—grudgingly allowed by a suspicious Stalin to be held in Moscow—had reviewed the worldwide evidence, and not least the Soviet Union's own Permian formations (which Murchison back in 1841 had named after the city of Perm near the Urals). In the 1950s Schindewolf studied a fuller sequence in the Salt Range, in the then new state of Pakistan, and concluded that the extinction must have been sudden, but he failed to persuade other scientists. In 1961, however, two American palaeontologists, Curt Teichert and Bernhard ("Bernie") Kummel, visited all the sites around the world where such sequences across the boundary were known. Subsequent fieldwork by others, at an increasing number of such sites, confirmed their conclusion that the extinction event had been both sudden in timing and drastic in character. (This necessarily international research was hampered for many years by the effects of the Cold War, since some of the most promising sequences were in the Soviet Union and China; but from the 1980s onwards Russian and Chinese geologists and palaeontologists collaborated increasingly with those from the West, to the great enrichment of their science.)

As the evidence for this mass extinction accumulated during the 1960s, those scientists who were convinced by it argued—although at that time to little effect—that the possible reality of the event should not be dismissed just because no adequate cause for it had yet been suggested (I was one of these, citing in my case the fossil record of brachiopods, a major group of invertebrate animals that was flourishing until the Permian period but never again developed any comparable diversity). The historical reality of a drastic moment or period of mass extinction at or around the Permian-Triassic boundary became in these years almost irrefutable. But it was only later, as the widespread hostility towards catastrophism began to subside, that possible causes for it were suggested. Summarized very crudely, in the 1970s explanations linked to the then newly acceptable ideas about shifting continents seemed most attractive; in the 1980s, those related to possible impacts of comets or asteroids from outer space; in the 1990s, those linked to the clear signs of exceptional volcanic mega-eruptions at just the time of the extinctions. By the early 21st century the balance of opinion among the relevant scientists was swinging in favor of a multi-causal but probably Earth-bound explanation. In contrast, for the second most drastic mass extinction event, at the end of the Cretaceous period—the boundary between the Mesozoic and Cenozoic eras—an extra-terrestrial explanation became the most popular and persuasive (and will be summarized later in this chapter). With mounting evidence for these two major episodes of mass extinction, and less unequivocally for at least three others, there was a dramatic change of outlook among Earth scientists: it was as striking as their belated acceptance of shifting continents and oceans in the theory of plate tectonics. In the late 20th century they dropped their rather rigid and often dogmatic interpretation of Lyellian "uniformity," and were prepared instead to consider the possibly major role that events far beyond human experience might have played in the deep history of the Earth and its life.

## UNRAVELING THE DEEPEST PAST

The geochronology made possible by radiometric dating had its most profound impact, however, in the furthest reaches of the Earth's history. Some of the very first radiometric dates to be calculated, such as Holmes's, confirmed what only a few geologists in the later 19th century had suspected on the basis of their fieldwork among the oldest rocks. To the surprise of many in the early 20th century, the whole of the well-charted sequence of geological periods from the Pleistocene back to the Cambrian was apparently dwarfed by a far longer pre-Cambrian or "*Precambrian*" history. Successive radiometric dates showed ever more conclusively that the Precambrian was far from having been a relatively brief prelude: perhaps no longer than, say, the Palaeozoic era or, at most, comparable to the time from the start of the Palaeozoic to the present. On the contrary, the Precambrian seemed to comprise by far the greater part of the entire history of the Earth. The new radiometric dating therefore demanded a radical change in geologists' perspective. This was signaled indirectly by the introduction in 1930 of the new term "*Phanerozoic*" ("evident life"), to denote the totality of the Palaeozoic, Mesozoic, and Cenozoic eras, or the Earth's entire history since the start of the Cambrian. Even this vast span of time—which was estimated on radiometric evidence at somewhere around 500 Ma—seemed to be dwarfed by a far lengthier Precambrian history, in which there were few if any unambiguously "evident" signs of life.

However, the almost complete failure of the fossil record when traced back into the Precambrian was no longer such a formidable barrier to understanding the early history of the Earth, for radiometric dating began to make it easier to identify sets of rocks of different ages *within* the Precambrian. Back in the later 19th century some geologists, using evidence from their fieldwork, had already distinguished between the *Archaean* rocks that formed the "basement" of most Precambrian regions, and the overlying *Proterozoic* formations that in some regions looked like possible sources of early signs of life (they were certainly Precambrian because they clearly underlay formations with Cambrian and other Palaeozoic fossils). Radiometric dating confirmed this distinction, for Archaean rocks yielded extremely ancient dates and Proterozoic formations somewhat less ancient ones. But it also began to make it possible to sort out the relative ages of different groups of formations *within* the Proterozoic era. In effect the well-tried methods of stratigraphy could be extended into the Precambrian history of the Earth, even without the help of the fossils that had proved so valuable in establishing its subsequent Phanerozoic history (this involved re-discovering or at least re-using what had been the methods of *geognosy*, before they were enriched by the use of fossils and transformed into William Smith's *stratigraphy*). Fieldwork assisted by the new radiometric dating showed that the vast expanses of Precambrian time had been just as much packed with eventful change as the Phanerozoic. For example, there had evidently been many successive episodes of orogenic (mountain-building) movements in different parts of the world in the course of the Precambrian; and by the end of the 20th century these were being interpreted in terms of the movements of very ancient tectonic plates. More surprisingly, ancient tillites were discovered among Precambrian rocks, which implied that there had been one or more Precambrian ice ages long before the Carboniferous ice age on the Gondwanaland continents, just as that ice age had happened long before the recent Pleistocene ice age on the northern continents.

This might have suggested a somewhat Lyellian view of an Earth in a steady state, or at least one in which similar events had occurred, and similar processes had been in play, as far back in the past as the evidence could be traced. But other evidence suggested long-term trends that might have made the early Earth a radically different kind of place from what it subsequently became. For example, in 1963 the American palaeontologist John Wells showed that the micro-structure of fossil corals—if preserved well enough to show their daily growth rings—indicated that, even as (relatively) recently as the middle of the Palaeozoic era, there had been about 400 days in the year; this was later confirmed by much other fossil evidence. The slowing down of the Earth's rotation was attributed to the effects of tidal friction. Extrapolated back into Precambrian times, it suggested that early in its history the Earth was probably spinning on its axis much more rapidly than at present, making days and nights much shorter. In such a fundamental way, the past "uniformity" of the Earth was apparently far from absolute.

An equally striking characteristic of the early Earth was inferred from the peculiar "*banded iron formations*" found in several Precambrian regions but never among Phanerozoic formations. Unlike later iron ores, they seemed chemically unlikely to have been deposited in waters rich in oxygen. This suggested that early Precambrian seas, and the atmosphere above them, might have been poor in oxygen or even lacking it altogether. A conference in 1965 on "The Evolution of the Earth's Atmosphere" signaled a newly explicit discussion of the previously implicit assumption—adopted by most geologists, again under the enduring influence of Lyell's "uniformity"—that a feature as fundamental as the atmosphere must have been more or less constant throughout the Earth's history (Adolphe Brongniart's conjecture, early in the 19th century, that its composition might have changed substantially even since the Carboniferous period, had been neglected or forgotten). By the late 20th century some geologists were claiming that the point at which free oxygen began to be added to the atmosphere had been one of the most significant moments in the entire history of the Earth: it was dated early in the Proterozoic era and named, perhaps somewhat over-dramatically, the "*great oxygenation event.*"

The vast expanses of Precambrian history were almost, but not quite, devoid of any fossil record. Ironically, the contrast with the subsequent Phanerozoic history of life was accentuated in 1909 by Walcott's discovery that the Burgess Shale, a formation found high in the Canadian Rockies, contained an astonishing assemblage of Cambrian fossils. Like other cases of exceptional deposits (*Lagerstätten*) preserving the "soft parts" of fossil organisms rather than just those with "hard parts" such as shells, it was an unexpected window onto the prolific life of the time; it was rather like the 18th-century excavations of Pompeii and Herculaneum, which had revealed the everyday life of the Roman world far more fully than the ordinary run of its ruined temples and theaters. The Burgess Shale was the earliest of these rare and valuable windows onto the deep past, and it showed that Cambrian seas had teemed with Cambrian life. And although it was not fully apparent until the later 20th century, when the Burgess fossils were studied more intensively and with greatly improved techniques, these Cambrian

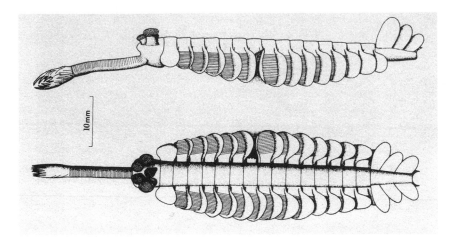

**FIG. 11.2** One of the more peculiar of the Cambrian fossil animals preserved in the famous Burgess Shale in the Canadian Rockies. Unlike the trilobites found in the same rock, the strange five-eyed *Opabinia* (shown here in side view and from above) had no "hard parts" or skeleton that would have been preserved under more normal conditions. This reconstruction—based on specimens squashed flat when the mud they were buried in was compacted into shale—was published in 1975 by Harry Whittington, the leader of the research group that in the late 20th century re-studied Walcott's earlier specimens and collected many more. The Burgess Shale highlighted the puzzle of the evolutionary origin of the quite large and diverse animals that made their (relatively) sudden appearance in the "Cambrian explosion" of life, since they seemed to have been preceded, in the Earth's lengthy Precambrian history, only by microscopic forms of life.

animals were not only complex but also far more diverse than anyone had anticipated. This heightened the problem of understanding what kinds of life might have preceded them in the Precambrian.

The dramatic expansion of Precambrian time, as shown by the new radiometric dates, might have been seen as vindicating Darwin's confident assumption that there had been ample time for the very slow evolution of all the diverse and complex organisms that made their apparently sudden appearance in the Cambrian period (including those now revealed so unexpectedly in the Burgess Shale). But the Precambrian formations still failed to yield any evidence for any such gradual evolution. The very rare scraps of alleged animal fossils that Walcott and others found in Precambrian rocks in the early 20th century were generally dismissed—like the notorious *Eozoon* back in the 19th century and with equally good reason—as either inorganic in origin or too dubious to count. And in the 1930s a leading palaeo-botanist, the Englishman Albert Seward, expressed similar scepticism about other alleged Precambrian fossils, such as Walcott's pillow-like *Cryptozoon* structures. His authoritative opinion discouraged others from continuing to search at all for Precambrian fossils. In 1953, however, the American geologist Stanley Tyler, while studying undoubtedly Precambrian rocks on the shores of Lake Superior, happened to find in the Gunflint formation a chert (a flint-like rock) in which there were abundant and well-preserved "*microfossils*," visible only with a microscope. Tyler showed them to the palaeo-botanist Elso Barghoorn of Harvard, who identified them as definitely organic filaments and spores. But the case remained for some time controversial and almost unique. Reports by Russian geologists that they had found other microfossils in Precambrian rocks in the Urals were met with scepticism by scientists in the West, partly from Cold War distrust and suspicion of the Soviet scientists' methods and standards. In 1965, however, Barghoorn's student Bill Schopf, while working primarily on the Gunflint fossils, found microfossils in a similar Precambrian chert in Australia, and other cases followed. He and the leading American

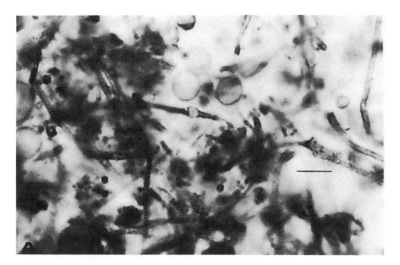

FIG. 11.3 Precambrian microfossils, first discovered in 1953 in the Gunflint formation (about 2000 Ma in age) on the north shore of Lake Superior. These spores and filaments were visible only when thin slices of the flinty rock were studied under a microscope; this portion of one such slice is only about 0.1 mm across. They proved that very small forms of life, such as bacteria, were present long before the evolutionary "Cambrian explosion" in which much larger animals were evolved.

palaeontologist Preston Cloud both concluded that the evidence now pointed to the possibility that the "Cambrian explosion" in the fossil record might reflect the relatively sudden appearance of *large* organisms (multicellular or "*metazoan*" animals such as trilobites), and that it might have been preceded by a far more lengthy history of exclusively microscopic forms of life.

This idea was modified but not negated by the discovery, in the remote Ediacara Hills in South Australia, of many large-sized but wholly "soft-bodied" fossils. They were just impressions on a rock surface, such as a jellyfish might leave if stranded on a modern seashore. In a story somewhat parallel to the Gunflint case, these fossils had first been found in 1943 by the Australian mining geologist Reg Sprigg; but they were generally dismissed, either as only questionably organic in origin or—because they were so large—as organisms probably of Cambrian or later date. It was not until the 1960s that the Austrian

FIG. 11.4 One of the strange late Precambrian fossils (*Dickinsonia*, about 6 cm across) from Ediacara in South Australia, preserved as a mere impression on a rock surface. In the 1950s this and the other "soft-bodied" fossils from Ediacara were accepted as the first known set of quite large animals, probably "metazoans," dating from shortly—in geological terms—before the first Cambrian shelly fossils, and representing an early phase in the "Cambrian explosion" of life. This photograph was published in 1961 by Martin Glaessner, who first made the fossils of the "Ediacaran period" widely known.

palaeontologist Martin Glaessner, who had immigrated to Australia and then focused his research on them, confirmed that they came from a formation well below another that contained characteristic early Cambrian fossils. This was later supported by radiometric dates that put the Ediacara fossils clearly in the Proterozoic era, though not long (by geological standards) before the start of the Cambrian period. These very strange forms of life were difficult or even, in some cases, impossible to assign to any known later group of animals, either fossil or living. Elsewhere, one similar specimen had been found in Precambrian rocks in Charnwood Forest in England as early as 1957, and others were later found in abundance in rocks of about the same age in several parts of the world, particularly at Mistaken Point on the coast of Newfoundland and in the Doushantou formation in south China. In 2004 they were all used to define a distinctive worldwide *Ediacaran period* immediately preceding the Cambrian.

The Ediacara fossils, considered in relation to the microfossils from much earlier in the Precambrian, suggested that the Cambrian "explosion" might have entailed *first* the evolution of large size and *then*, somewhat later, the evolution (among some but not all kinds of animals) of the hard shells that can be preserved as fossils under more ordinary circumstances. From the 1970s onwards this second phase was clarified by detailed studies of early Cambrian formations, notably in some remote parts of Siberia and China (here again the work of many fine Russian and Chinese scientists, as the Soviet Union disintegrated and China recovered from its disastrous Cultural Revolution, was of great importance). This fieldwork showed that the earliest animals with shells of any kind—those likely to be preserved under more ordinary conditions than the Burgess or Ediacara cases—had not appeared all at once but quite gradually. First, in the lowest and very oldest Cambrian formations, there were only some small shelly fossils, formed by organisms of uncertain kinds; then came formations with fossils recognizably similar to some modern animals (brachiopods); and these were later joined by the first trilobites, which became larger and more diverse during the rest of the Cambrian period. This suggested that shells might have been acquired by one kind of animal after another, spread over a relatively long span of time. As usual, the causal questions about this were distinct from the historical sequence itself. Among other possible causes, there was debate about whether the evolution of shells had been triggered by changes in the composition of sea water, which might for the first time have allowed organisms to secrete the shelly materials, or whether it had been an evolutionary response to the appearance of the first predators.

However, in the perspective of the immensely long Precambrian history of the Earth, the first appearance of the diverse and complex Cambrian animals still seemed relatively sudden. It was an event or sequence of events that could still, with excusable exaggeration, be called a Cambrian *explosion* (even if its earliest phase, in the Ediacaran period, was now formally assigned to the very late Precambrian). But what also began to emerge in the 1960s was the possibility that the Ediacaran phase had itself been preceded by an exceptionally catastrophic climatic event or sequence of events. Precambrian tillites, and the ice ages they might represent, had already been under discussion at the International Geological Congress back in 1937. But this puzzle took a new turn when in 1964 Harland claimed that in late Precambrian formations in or near the present Arctic regions there were tillites that on palaeo-magnetic evidence must have been deposited *near the equator* of the time. This implied not only that the relevant landmasses had changed dramatically in latitude since the Precambrian—Harland was among those in Europe who had long been convinced of the reality of "continental drift"—but also that any late Precambrian ice age must have been far more severe than either of the two ice ages known from subsequent Phanerozoic history. Ice sheets, or at least floating icebergs carrying the erratic blocks now embedded in the tillites, must have extended over most or all of the globe. One or more such episodes of almost global glacial conditions had apparently been followed fairly soon—as usual, in geologists' conception of soonness—by the Cambrian explosion of life. This suggested a possible causal connection between these two equally dramatic events. A global ice age would certainly have disrupted previous environments catastrophically; and its aftermath, when the Earth warmed up again, might have offered exceptional opportunities for new forms of life to proliferate (my own brief published suggestion to this effect was combined with Harland's weighty case for a near-global ice age). But these ideas gained little support until, in the late 1990s, Paul Hoffman of Harvard and his colleagues argued, on the basis of further fieldwork around the world, that several such "*Snowball Earth*" episodes had preceded the Edia-

FIG. 11.5 A late Precambrian tillite in Norway: one of the photographs published by Brian Harland and myself in 1964 to support our claim that these clear signs of ice action, combined with palaeo-magnetic evidence that this part of Europe was then *near the equator*, implied that in late Precambrian time the planet had been a *"Snowball Earth"* (as it was later called). Any geologist seeing these photographs—without visiting the place and finding the tillite to be a solid rock and certainly Precambrian—would take it to be an ordinary till or "boulder clay" dating from the quite recent Ice Age of the Pleistocene period; it was resting on a surface of even older rock (on which the hammer is placed) marked by deep scratches likewise indistinguishable from those caused by similar stones embedded in Pleistocene and modern glaciers.

caran period towards the end of Precambrian history. There was also evidence that the level of free oxygen in the atmosphere had risen rapidly around the same time, which might have allowed the evolution of relatively large animals such as those at Ediacara and in the early Cambrian.

Before this eventful prelude to the Cambrian explosion, which might have kick-started the evolution of the large and "evident" life that extended all through the Phanerozoic to the present, the vast tracts of much deeper Precambrian time gradually yielded a better record of the even earlier history of life. Fossils representing microscopic organisms of various kinds were found in more and more Proterozoic rocks, some younger than the Gunflint chert but others even older. Different in character, but equally revealing, was the reha-

FIG. 11.6 A Precambrian stromatolite: a pillow-shaped mound of limestone, naturally broken open and revealing the banding that shows how it grew in size by gradual accumulation. This example, from a Proterozoic formation in the Grand Canyon, was published by Preston Cloud in 1988 to illustrate his interpretation of the history of life, according to which microscopic forms of life, such as the "cyanobacteria" that still produce stromatolites in the present world, existed long before any larger forms of life evolved. The fossil record of stromatolites pushed the origin of life itself back to a quite early point in the total history of the Earth.

bilitation of Walcott's pillow-sized *Cryptozoon* structures, which came to be accepted as genuine records of an important early form of life. These "*stromatolites*" ("rocky pillows") had been found in formations of many different ages, mostly Precambrian but also less commonly Phanerozoic. They were now interpreted as the products of microscopic life, forming "*microbial mats*," secreting or trapping mineral material and therefore growing slowly upwards to form large mounds. This interpretation was confirmed sensationally when in 1954 oil geologists found by chance that modern stromatolites were being formed in a salty lagoon at Shark Bay on the coast of Western Australia. It was perhaps the most important kind of "living fossil" ever discovered. By the end of the 20th century stromatolites were known from as far back as the Archaean era, and some were being dated at about 3500 Ma. Since the microscopic organisms building the modern stromatolites were confirmed as being very simple forms of life (*prokaryotes*), these ancient stromatolites implied that life—at least of this kind—was already present quite soon (relatively) after the Earth was formed as a planet. This fossil evidence contributed to a continuing lively debate among biologists about the even earlier origins of life itself.

It was equally significant that the modern stromatolites were found to be due more specifically to microscopic "blue-green algae" (*cyanobacteria*) that live by photosynthesis. Like modern plants, including ordinary algae or seaweeds, these get their energy from sunlight and produce oxygen as a waste product. This suggested a possible link with the evidence that the Earth's early seas and atmosphere had lacked oxygen altogether. The "great oxygenation event" early in the Proterozoic era might represent the point at which the continuous production of oxygen *by living organisms* had reached a level at which free oxygen could begin to accumulate in the waters of the Earth's seas and in its atmosphere above. This alone would have made possible the later evolution of all those other organisms—including of course eventually ourselves—that require oxygen for their life processes. In this way the history of life might have been integrated with the history of the Earth itself, far more intimately than had previously been imagined, as an "Earth *system*" (to which, controversially, James Lovelock's "*Gaia hypothesis*" of the 1970s attributed self-regulating organism-like properties).

## THE EARTH IN COSMIC CONTEXT

In the light of all this, geologists felt encouraged to make more searching comparisons between the Earth and the other bodies in the Solar System. It made them think about the Earth in a cosmic context, rather than as a body more or less insulated from any events or processes outside itself. This was a new direction for the Earth sciences to take, or rather, a revival of a much older way of thinking. Back in the 17th and 18th centuries, a close relation between the Earth and the rest of the cosmos had been taken for granted, particularly in the genre of "theory of the Earth": Descartes' and Buffon's theories were influential examples. But in the early 19th century most geologists had firmly rejected this kind of speculative theorizing. They had focused their attention instead on what could be observed directly, on strictly terrestrial events and processes, and they had neglected the cosmic dimension (De la Beche has been mentioned as a rare exception). The increasing differentiation of all the sciences in the course of the 19th century had heightened this effect. Each group of "men of science" was well aware of what those in other groups were doing, and cultivated friendly relationships with them, but they were in effect staking out distinctive intellectual territories over which they could properly claim authority and expertise. So, for example, the sciences of the Earth, even in the later 19th century, were rarely linked to those concerned with the extra-terrestrial universe. Kelvin's dogmatic assertion that physics and cosmology set strict limits to the Earth's timescale was an intrusion into geology that was unwelcome and ultimately rejected by geologists, and with good reason.

One of the few positive exceptions to this, in the later 19th century, had been Croll's idea about a possible astronomical cause for the sequence of glacial and interglacial phases during the long Pleistocene ice age. This idea was revived and improved in the early 20th century, but it was still an unusual exception. In 1930 the Serbian astronomer Milutin Milanković calculated the effects of three known variables in the Earth's orbit around the Sun (eccentricity, axial tilt, and precession); in combination they could produce what came to be called "*Milankovitch cycles*" in the Earth's climate. But there were problems with the idea, and it was not until 1976 that a key Anglo-American paper, titled "Variations in the Earth's Orbit: Pacemaker of the Ice Ages," marked the start of a more satisfactory integration of geology and astronomy on this issue. A combination of radiometric dating and isotopic evidence for palaeo-climates—derived from cores of sediment retrieved from the ocean floor and cores of ice retrieved from the Earth's remaining great ice sheets—confirmed the basic validity of the Milankovitch cycles; they were found to have had their strongest effect at a frequency of about 100,000 years. But although they came to be accepted as an important factor underlying the fluctuating climate of the Pleistocene period, they were evidently not the whole causal story, which seemed to be far more complex. They hardly lessened the sheer contingency of the geologically recent history of the Earth's climate (and therefore also of its likely future).

Another sign of the possible impact—in more senses than one—of the extra-terrestrial cosmos was the enigmatic event that in 1908 flattened a vast area of uninhabited forest in Siberia and lit up the sky from far away. Owing to the political turbulence in Russia and the remoteness of the site, it was not until 1927 that scientists studied on the spot what might have happened in the "*Tunguska event.*" Surprisingly, there was no crater and no clear trace of any large meteorite, but something might have exploded high in the atmosphere, perhaps vaporizing almost completely but producing a violent shock wave on the ground (later research suggested it had been comparable in magnitude to one of the first thermonuclear explosions). Whether it had been a rocky asteroid or an icy comet was much debated, but in either case the Tunguska event showed that intrusions from elsewhere in the Solar System were not limited to the quite frequently witnessed falls of small meteorites. Yet geologists remained reluctant to concede that major "catastrophic" impacts from space might have been a significant feature in the Earth's past history: they continued in practice to treat the Earth as a closed system almost insulated from the rest of the Solar System.

However, a similar impact event in the more remote past (though still very recent by geological standards) was suggested by a strikingly well-preserved crater, over a kilometer in diameter, that had been discovered in a remote desert area of Arizona. In 1891 Grove Karl Gilbert, the then head of the US Geological Survey, had concluded—in line with the usual assumptions among geologists at that time—that it was probably caused by an underground volcanic explosion and not by any kind of impact. Other American geologists had accepted his verdict. Nevertheless, in 1903 the mining entrepreneur Daniel Barringer staked a claim, literally, that a huge and hugely valuable iron meteorite might be buried beneath it. His project was a commercial failure, for nothing but small pieces of meteoritic iron were found. Not until 1960 was his interpretation fully vindicated (and the modern name "Meteor Crater" justified), when the American geologist Gene Shoemaker reported at an International Geological Congress in Copenhagen that he had found a peculiar variety of quartz (*coesite*)—previously known only from experiments in a laboratory—not only in the rocks around the crater but also on the test site of a nuclear explosion in Nevada. Since this distinctive "*shocked quartz*" was apparently formed only under conditions of extremely high pressure, it could be treated as a reliable "fossil" trace or marker of genuine impacts from outer space.

**FIG. 11.7** Meteor Crater (formerly known as Barringer Crater) in Arizona, about a kilometre in diameter. Aerial views such as this, which show its striking similarity to lunar craters, were not available when its origin—either as an impact site or as volcanic—was first debated by geologists around the end of the 19th century and the start of the 20th. On the floor of the crater are traces of the mine workings made during Barringer's fruitless search for a buried giant meteorite.

Shoemaker then found the same key mineral in the far larger Ries Crater in Bavaria, some 24 km in diameter, which was less well preserved and on geological evidence much older (dating from the Miocene epoch in the Cenozoic era); this supported what another American geologist, Robert Dietz, had already inferred after finding another distinctive high-pressure structure ("*shatter-cones*") in the rocks forming its rim. Dietz was convinced that there were many other impact craters around the world, being either mistakenly attributed to volcanic action or too much eroded to be easily recognized as former craters; he called them all "*astroblemes*" ("star-wounds"). In 1961, for example, he interpreted the Vredefort Ring, a very large circular geological structure in South Africa, as the deeply eroded remnants of a huge crater probably of Precambrian age. And in fact the Canadian astronomer Carlyle Beals had already initiated a systematic aerial search for such structures in the ancient rocks of the vast Canadian Shield, and by 1965 more than twenty had been identified there.

This suggested that there really had been occasional major impacts from asteroids or comets throughout the Earth's history. Yet this was a conclusion that most geologists were still reluctant to accept. What changed their collective mind was the parallel case of the craters on the Moon. Lunar craters, which were of course known only from telescopic observation, had long been regarded as extinct volcanoes. Only a minority of astronomers—and even fewer geologists, though Wegener was among them—had argued that they might be impact craters. In the 1960s Shoemaker was still in a minority on this point, and he regarded his extension of the idea to terrestrial craters as quite heretical. In any case the Moon had received little attention of any kind from astronomers in the earlier 20th century. The Moon and the planets were considered far less glamorous subjects for research than the Sun and other stars; most exciting of all—and transforming the science of cosmology at this time—were the nebulae that were newly recognized as other galaxies far beyond the Milky Way, in an ever-expanding universe far more vast than anyone had previously imagined.

What dramatically changed this situation, and made our nearest cosmic neighbor the focus of close scientific attention, was the American space program, which was a Cold War reaction to the launch in 1957 of the Soviet Sputnik, the world's first artificial satellite. But the later American Apollo missions to the Moon might have had little or no input from geologists, or any impact on the Earth sciences, if Shoemaker had not lobbied hard for the inclusion in the space program of what he called "*astrogeology*" (a term no less appropriate than "*astronaut*," though neither involved reaching literally for the *stars!*). The Moon was mapped in much greater detail than ever before, in preparation for the first manned lunar landing in 1969. A "relative" chronology was worked out for the Moon, using methods borrowed from terrestrial stratigraphy (for example, craters of later date truncated earlier ones, like unconformities between formations of strata); lunar history was reconstructed, and divided into named eras that paralleled those of the Earth's history. The manned landings then provided rock samples that settled the origin of the lunar craters in favor of impacts, and also enough material for an outline of an "absolute" chronology based on radiometric dating. The Moon turned out to be of about the same age as the Earth (and most meteorites). There seemed to have been a major *"heavy bombardment"* episode by large asteroids or similar bodies very early in its history, and smaller and less frequent impacts ever since ("bombardment" was an appropriate term, since the lunar craters had the same physical character as the much smaller craters produced by wartime bombs). In the absence of any significant erosion or atmosphere on the Moon, even the oldest lunar craters, equivalent in age to the earliest Precambrian, were still well preserved.

This prompted a much more thorough comparison with the Earth's history. The rarity of impact craters on the Earth, compared with the Moon's pockmarked surface, no longer seemed surprising: the Earth's atmosphere, its far more active processes of erosion, the vast extent of its oceans, and the partial destruction of its ancient continents by the processes of plate tectonics, were more than enough to explain the contrast. But in fact by the 1980s more than 200 terrestrial impact sites or astroblemes had been identified: either relatively well preserved craters such as the Barringer and Ries, or at least the circular structures such as the Vredefort that were interpreted as their deeply eroded "fossil" remains. And by analogy with the Moon, the Earth's history could now be extended back into the deepest Precambrian time, before even the Archaean era: the Earth must have suffered the same "heavy bombardment" as the Moon, although no direct traces of it were known to have survived, during what Cloud in 1972 named appropriately a *"Hadean"* era (from the unquestionably unattractive hell of Greek myth). By the end of the century it was conjectured that this bombardment might have been the route by which the Earth first acquired the most essential component—water—that later made any kind of life possible.

In the 1970s most scientists came to accept that the Earth must have been struck by cosmic bodies of all sizes—from frequent falls of small meteorites to occasional impacts by massive comets or asteroids—throughout its history. This encouraged geologists to consider the largest of these events as possible causes for what they now recognized as its occasional major catastrophes, and particularly its episodes of apparent mass extinction. Such ideas were very much "in the air," but it was one specific claim that in 1980 brought the issue to the attention of a wider range of scientists and then of the general

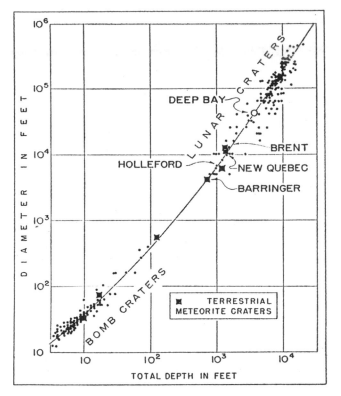

FIG. 11.8  A graph published by Carlyle Beals and his colleagues in 1963, helping to make the case for the reality of impact craters on Earth. The data on bomb craters and lunar craters had been published in 1949, as part of the argument that most of the Moon's craters were not volcanic in origin but the result of meteorite or asteroid impacts. Here the data on terrestrial craters were added, to support the claim that they too were caused by impacts. Apart from the well-known and geologically recent "Barringer" (or "Meteor") Crater in Arizona, the terrestrial craters marked here by name were among those that had recently been discovered in Canada. The logarithmic scales brought together craters across a vast range of sizes, from a few meters to around 200 km in diameter, all of a similar form.

public. The prominent American physicist Luiz Alvarez, collaborating with his geologist son Walter and other colleagues, reported a "spike" in the concentration of the rare element iridium, in a thin layer that marked the "*K/T boundary*" between the youngest Cretaceous (*Kreide* [*Chalk* in German]) and the oldest *Tertiary* (Cenozoic) ones, in a sequence of strata near Gubbio in northern Italy. These were rocks that had been deposited in deep water, and the only fossils found close to the boundary were those of microscopic organisms. But the report by the Alvarez group entailed the much more sweeping claim that the probable cause of the mass extinction that brought the Mesozoic era to an end was nothing less than the devastating impact of a large asteroid (like some meteorites it might have had a telltale iridium content much higher than in terrestrial rocks). This was sensational enough, but it became much more so when it was popularized by the media into a lurid melodrama of the sudden mass killing of all the Mesozoic dinosaurs, which were still—a century and a half after their first discovery—the public's favorite fossil wildlife.

Those best able to judge the matter were more sceptical. In particular, the experts on dinosaurs argued—from the detailed fossil record of dinosaurs around the world—that this whole group of reptiles had already been in slow decline through ordinary piecemeal extinctions, long before the end of the Cretaceous period, and that only the last few stragglers might have been wiped out around that time. Yet the prestige of a Nobel prizewinning physicist, combined with publication in the leading scientific periodical *Science*, and above all the deployment of highly sophisticated laboratory techniques, certainly helped support the case for a cosmic catastrophe. The sceptics found themselves in a minority. As often happens in scientific controversies, however, the argument provoked a search for better evidence on both sides. In this case it was the impact theory that on balance gained further support. Further signs of an exceptional event were found at the K/T boundary at widely scattered places around the world: not only an iridium "spike" but also, for example, plausible traces of the mega-tsunami and widespread forest fires that such a huge impact would almost certainly have caused. The absence of any obvious impact site of the expected size was a serious weakness. But in 1991 a huge circular structure on the Yucatán peninsula in Mexico, deeply buried beneath Cenozoic sediments and detectible only by geophysical methods, was widely agreed to be the likely missing site. However, even Chicxulub (as the buried crater was called, from one of the villages above it) did not silence the sceptics, and in the early 21st century the conclusion reached by many geologists was that the intruder from outer space might have been a "last straw" piled on top of an environmental crisis already under way for other unrelated reasons.

The increasing acceptance of an impact from outer space, as at least a major factor in the K/T mass extinction if not the sole one, led some geologists to try to apply a similar explanation to the even greater event at the Permian-Triassic boundary and to the other suspected episodes of mass extinction in the Earth's history. Raup and Sepkoski, whose quantitative analysis of the whole Phanerozoic fossil record had helped to identify these points, went still further when in 1984 they suggested that such events might have happened with periodic regularity, about every 26 Ma. A hypothetical nearby star that they named Nemesis, if coupled with the Sun in some kind of double-star system, might have periodically perturbed the orbits of comets on the outermost fringes of the Solar System, which could have greatly heightened the chance that one such comet would be thrown off course and collide with the Earth. Although this Nemesis theory did not gain wide support, it did illustrate how thoroughly, by the late 20th century, geologists had absorbed the cosmic dimension into their thinking about the Earth. It was rather appropriate that in 1994 Shoemaker's vision of the Earth's continued interaction with the rest of the Solar System was illustrated spectacularly, when the comet "Shoemaker-Levy 9" (named after him, his wife, and another colleague) was closely watched by the world's astronomers as it crashed predictably onto Jupiter with huge impact effects (although, since Jupiter is not a rocky planet but a "gas giant," it left no permanent trace). Catastrophic cosmic impacts, on a far larger scale than the Tunguska event, were seen to be a genuine "actual cause," or what Lyell had defined as an observable "cause now in operation."

While the fallout from the space program led geologists to adopt a fully cosmic perspective on Planet Earth, the compliment was returned during the same decades. Using the Earth's deep history as a model, astronomers increasingly interpreted the other bodies in the Solar System in terms of their "*planetary histories*." Whatever in principle they may previously have thought about, say, Mars or Venus, in practice they had generally treated the planets and their satellites as objects with accurately known orbits and other physical properties, but without any knowable *histories* of their own, apart from their inferred involvement in the distant origin of the Solar System itself. However, once the Moon had been thoroughly historicized by the application of "astrogeology" to it, both before and after the lunar landings, it was a short step—at least conceptually—to apply the same kind of interpretation to what was revealed by the unmanned missions to

**FIG. 11.9** Planet Earth seen from space: the celebrated "Blue Marble" photograph taken in 1972 from the manned Apollo 17 spacecraft on its way to the Moon; it showed the whole of Africa flanked by the South Atlantic and Indian Oceans, with the Arabian peninsula above and Antarctica below, and with clouds and swirling storm systems. This and other similar images had a profound impact on the public understanding of the Earth as a precarious "ball in space." For scientists, however, it also gave, more specifically, a vivid impression of the Earth as a complex but unified system—solid, liquid, and gaseous (lithosphere, hydrosphere, and atmosphere)—at the present moment in its immensely lengthy history and likely future. It stimulated comparison with the other planets and their satellites, with their equally lengthy but widely differing "planetary histories."

the planets. So, for example, Bretz's channeled scabland in Washington State was found to be replicated on a much larger scale on Mars, giving evidence that in its deep past that planet had had plenty of water at its surface, although since that time it had become a dry desert. Conversely, the decidedly Hadean conditions discovered on the surface of Venus below its thick cloud cover could be interpreted in terms of a planetary history that was radically different from that of either the Earth or Mars, yet one that might have had a distant starting point in common with theirs. The discovery that among Jupiter's Moon-sized satellites Europa is wholly covered in a thick sheet of ice—a small-scale equivalent of a Snowball Earth—while Io is studded with active volcanoes, just added to the unexpected variety of the bodies in the Solar System, and the equally diverse histories that must have brought them into being.

In the light of this novel cosmic perspective, the history of the Earth itself was re-conceived in the later 20th century and early 21st as one specific case in a much wider set of divergent planetary histories: one specific pathway of change among others, each probably as individual and contingent as the rest. This focused attention on the very specific circumstances that had underlain the Earth's particular history—for example, being neither too near the Sun nor too far away—which in turn governed the astronomers' search for indirect evidence of "*exoplanets*" orbiting other stars (the first was reported in 1992) and their estimates of how many of these might be rocky or even Earth-like. This was combined with biologists' conjectures about all the further circumstances that might enable, or limit, the generation of any kind of life on other planets. In this context the evolution of highly complex forms of *intelligent* life, which might give substance to earlier speculations about a possible "plurality of worlds," became more constrained and improbable than ever. That one of the rocky planets orbiting our local star had nonetheless become the abode of living organisms—and eventually of intelligent beings capable of discovering and reconstructing its past history with some confidence and reliability—was just the most remarkable of all these complex contingencies.

# 12

# Conclusion

By the early 21st century the Earth's particular planetary history had been reconstructed in impressive detail, and had been found to be surprisingly eventful throughout. The main outlines of the Earth's deep history had become uncontroversial, at least in the sense of establishing the correct sequence of distinctive periods and notable events, if not of working out all their underlying causes. Once geologists had recognized that their geological periods and other named spans of time were matters of convention and convenience, any arguments about their definition could be, and generally were, settled by discussion and negotiation. A hierarchy of time-spans—*aeons, eras, periods, epochs*, and still briefer units—had been agreed upon as useful for describing and explaining the broad features of the Earth's history and also its details. They were invaluable for describing this lengthy and eventful history, despite not being, at least in the first instance, dated in years.

During the 20th century the Phanerozoic *aeon*, with its fairly complete and continuous fossil record, was recognized as just the most recent major part of a history that extended back into no fewer than three vast earlier aeons (Proterozoic, Archaean, and Hadean) with a far more sparse fossil record or none at all. Within the Phanerozoic aeon, the three great *eras* in life's history (Cenozoic, Mesozoic, and Palaeozoic) that Phillips had named in the 19th century were recognized in the late 20th as being separated by the two greatest episodes of mass extinction. Within each of Phillips's eras, the *periods* that were such impressive products of 19th-century stratigraphy (such as Murchison's Silurian) had been given agreed-upon definitions that made them invaluable for tracing the history of the Earth and its life, with its changing environments, shifting continents, and occasional crises and catastrophes. And the still finer divisions of time and history represented by *epochs* (such as the Pleistocene and Eocene that Lyell in the 19th century had defined within the Cenozoic era) had likewise proved their worth in reconstructing the Earth's history in greater detail.

This dividing up of the Earth's history was still open to further refinement. Early in the 21st century, for example, some Earth scientists proposed adding to Lyell's epochs by distinguishing a new and still current *anthropocene* ("humanly recent") epoch. This recognized the profound *material* impact that the human species has already had on the Earth, during the geologically minuscule time-span since the start of the Industrial Revolution. For example, the modern world's detritus of throwaway plastics, which has become so obtrusive on even the most remote shorelines, would certainly provide an exemplary "characteristic fossil"—new and distinctive, sudden in appearance, and worldwide in distribution—for any stratigraphers in the far distant future, identifying the sediments or strata formed during the past hundred years. More seriously, the rapidly escalating numbers and increasingly disruptive environmental impact of the human species seemed to be causing a major episode of mass extinction, threatening to match in magnitude the five greatest ones that had been detected in the fossil record, to become the sixth major mass extinction in the Earth's Phanerozoic history. And the geologically sudden release of vast amounts of carbon dioxide into the Earth's atmosphere, as the result of the massive combustion of fossil fuels previously locked up underground for tens or hundreds of millions of years, was widely considered likely to be having equally sudden, profound, and lasting effects, which could hardly fail to to be accentuated in the near future. While there was every reason to suppose that the Earth's physical fu-

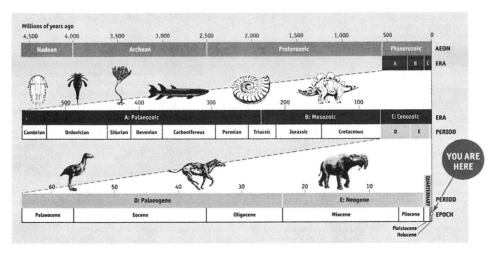

**FIG. 12.1** The Earth's deep history, on three different scales marked in millions of years before the present, showing some of its named spans of time. This diagram, designed in 2011 for the intelligent worldwide public that reads *The Economist*, summarized the whole eventful history of the Earth as scientists had reconstructed it by the early 21st century. It could be given its vast quantitative scale—with ever-increasing reliability and precision—thanks to a century of progressive improvement in the techniques of radiometric dating. For Phanerozoic time since the start of the Cambrian period (middle and lower timescales) the diagram also pictured a few distinctive animals, from trilobites to elephant-like mammals. It could also have marked (on the upper timescale) some of the likely landmarks in the still earlier or Precambrian history of the Earth, such as the cosmic "heavy bombardment" in the Hadean, the first microscopic life forms in the Archaean, the beginnings of an atmosphere with oxygen early in the Proterozoic, and the "Cambrian explosion" of relatively large organisms at the start of the Phanerozoic. But the diagram was designed primarily to explain the then recent proposal that the "epochs" of the Cenozoic era (lower timescale), as defined mainly by Lyell in the early 19th century, should be augmented with an "*Anthropocene*" epoch, too brief to be depicted here except by the signpost "You are here." This denoted the period since the start of the Industrial Revolution, during which the *material* impact of the human species had arguably become as far-reaching as, say, that of the huge climatic changes during the great Ice Age in the Pleistocene epoch.

ture *as a planet* would be as lengthy and eventful as its deep past, it seemed less certain that it would in the long run remain habitable by *Homo sapiens*. Some of the scientists who were in the best position to judge the matter even considered it possible that, unless the human species mends its ways in the very near future, the 21st century might be humanity's last.

From the early 20th century onwards, all this qualitative deep *history* of the Earth had been enriched by being calibrated against a quantitative scale of deep *time*. By the early 21st century, a hundred years of technical improvement—generating ever-increasing precision, reliability, and consistency—had turned the radiometric dating of minerals and rocks into a matter of routine. Not just the total age of the Earth but also, far more importantly, the dating of the complex sequence of events throughout its history, had become uncontroversial except on points of relative detail. Nor was this geochronology wholly dependent on the physicists' assumption that the rates of decay of radioactive isotopes, like other basic properties of matter, have remained constant through time. Other and independent methods of dating, such as the analysis of annual layers in sediments (varves) and in ice cores, had confirmed—at least for the more recent history— that the estimated orders of magnitude were correct. They had proved beyond reasonable doubt that thousands of years must have elapsed since even the end of the obviously very lengthy Pleistocene ice ages, which in turn were unquestionably just at the tail end of the Earth's total history. So a radiometric figure of several billions of years since the origin of the planet, rather than the mere tens of millions that the physicist Kelvin had allowed shortly before the discovery of radioactivity, seemed proportionate and consistent. Beyond reasonable doubt, the Earth was ancient to a degree that was literally almost inconceivable (and the universe even more so, as cosmologists had concluded on largely independent grounds).

The single most striking feature of this almost unimaginably lengthy history of the Earth—and not just because we ourselves are among its products—was the history of life. That life had had any true history, rather than remaining more or less the same all along, had been utterly uncertain until Cuvier and others in the early 19th century demonstrated the reality of extinction, which showed that the life of earlier periods had been distinctly different from that of the present world. His successors later in that century had then established, from the increasingly well explored fossil record, that the history of life was not only linear and directional but also in some sense "progressive." Forms of life that biologists regarded as "higher" or more complex (such as mammals) had generally made their appearance later in the fossil record than those that were simpler or "lower" (such as fish); Lyell's rearguard denial of this, claiming that the record supported instead a uniformitarian or Hutton-like "steady-state" interpretation of the world, had lost all plausibility. The apparent absence of any genuine human fossils, and then the discovery in the mid-19th century that they were present in the fossil record but confined to its most recent (Quaternary) part, just confirmed that the human species had only made its appearance at—relatively—the last moment.

At the other end of the fossil record, the discovery in the later 19th century that Cambrian rocks contained the remains of organisms almost as diverse and complex as those of later periods, and yet were underlain by Precambrian rocks with no unambiguous fossils at all, had left the beginnings of the history of life extremely puzzling. Early in the 20th century, however, the development of radiometric dating had radically altered the perspective of geologists and palaeontologists on this point: not so much by expanding the total age of the Earth, but much more profoundly by expanding the Precambrian into by far the greater part of the Earth's total history. Yet its fossil record had remained sparse and problematic until, in the later 20th century, new discoveries made it seem likely that living things had almost all been microscopic in size and relatively simple in structure throughout Precambrian time; only with the "Cambrian explosion" around the start of the Cambrian period had larger and more complex (metazoan) forms of life proliferated. What was equally unexpected was the discovery that life, at least in small and relatively simple forms, had made its first appearance quite early in the Earth's total history (in the Archaean). Also surprising was the evidence that it might later have been responsible for adding oxygen to the Earth's atmosphere, which could eventually have made possible the development of much more complex forms of life, including of course ourselves.

## PAST EVENTS AND THEIR CAUSES

Behind all this unexpectedly complex *history* of life, as reconstructed from the fossil record, lay the quite separate *causal* question of how its changes had come about. Many earlier naturalists had conjectured that later forms of life must have been derived from earlier ones, or perhaps directly from non-living matter, by natural processes of some kind; but what these processes had been was profoundly obscure (that new species had appeared by acts of direct divine intervention was *not* a common opinion among those, whether religious or not, who would later call themselves scientists). Darwin had claimed that his own particular version of evolution—imperceptibly slow, and driven mainly by natural selection—was the only truly scientific one available, and he had marshaled so much evidence in its favor that evolution had sometimes come to be equated with its "Darwinian" version. But in fact a wide variety of other kinds of evolutionary theory had been around at every stage of the debate, ranging, for example, from the "Lamarckian" in the 19th century to the occasional relatively rapid changes known as "punctuated equilibrium" in the late 20th: all of them were as natural in character as the Darwinian and were often considered equally good, or better, at explaining the evidence. However, geologists and palaeontologists could contribute little to these sometimes vehement biological arguments about the causal processes that might underlie evolutionary change. On the other hand they could and did insist that only *fossil* evidence—however imperfect the fossil record—could provide reconstructions of the course of evolutionary change that had some plausible grounding in historical reality. This remained the case even when, in the late 20th century, some of the conventional evidence for evolution (such as that of anatomy and physiology) was augmented by that of genetics and DNA sequences: this was still evidence based almost wholly on life as it is in the present evolutionary moment, not on what it might have been in the deep past. Anyway, the earliest fossil evidence was of very simple forms of life (*prokaryotes*) like modern bacteria; but since even these are, on a micro-level, already amazingly complex, palaeontologists could contribute little to the biologists' lively debates about the still earlier origin of life itself.

This distinction between the historical evidence for evolution (sometimes misleadingly called "evolution as a *fact*") and its causal explanation ("evolution as a *theory*") is just one example of what has emerged repeatedly during the discovery of the Earth's deep history. Establishing the historical reality of *any* event in the deep past—not only those involving living organisms—has always been distinct from finding an adequate causal explanation for it. Again and again in the course of geological research, the reality of a past event, or series of events, has been established well before its cause or causes were fully understood; again and again, those arguing that the events did happen have had to insist that the lack of a convincing cause for them was no reason to deny their historical reality. Obvious examples from the more recent history covered in this narrative have been the enigmas of the ice ages in the Pleistocene period, the gigantic nappes and overfolds in the Alps and other mountain ranges, and above all the movement of whole continents around the globe.

In all such cases, the distinction between establishing historical realities and finding causal explanations has been underlain by the difference between sciences such as history (using "sciences" in its original sense, still preserved in the German word *Wissenschaften*) and sciences such as physics (just one of the many sciences of nature or *Naturwissenschaften*). Recognizing the sheer diversity of the sciences, both natural and human, liberates them all from the straitjacket imposed by the false assumption—bolstered by the "Anglophone heresy" of a singular "Science"—that they all share, or should share, one unique "scientific method." The history of the discovery of the Earth's deep history has illustrated the crucial role played by the importation into the study of the natural world of the dimension of history itself, which was already well established in the study of the human world. Methods and concepts used in the study of human history were transposed into the study of the Earth and its features large and small. Mountains and volcanoes, rocks and fossils, were recognized as being the products of nature's *history*; they could not be understood solely in terms of *causes* governed by the timeless "laws of nature." And as in human history, the past events that could be inferred from features preserved in the present were marked throughout by contingency: they could not be predicted, even in retrospect (in technical jargon, retrodicted). At every point in any sequence of past events, it was always conceivable or imaginable that things might have happened differently, with a different train of consequences, without in any way violating the constant "laws of nature." Counterfactual or "what if?" history was always possible and often illuminating: as much to imagine what might have happened if the Earth had never been hit by the asteroid that supposedly wiped out the last dinosaurs 65 million years ago, as if the Austrian archduke visiting Sarajevo in 1914 had not been hit by the assassin's bullet that precipitated the First World War.

This element of sheer contingency has also permeated the process by which savants, naturalists, "men of science," or scientists—some of their successive designations—have pieced together what really did happen in the course of the Earth's deep history. As this narrative has emphasized repeatedly, discoveries of new evidence and the formulation of new and persuasive interpretations have been, again and again,

**FIG. 12.2**  A reconstructed scene from the Earth's deep history, as imagined by the Austrian palaeontologist Franz Unger and published in *The Primitive World* (*Die Urwelt*, 1851). It shows a scene in the Mesozoic era (more precisely the early Cretaceous period), with two *Iguanodon*—among the first fossil reptiles to have been defined as *dinosaurs*—competing for a mate, in an environment with similarly unmodern plant life (on which Unger himself was an expert). It was one of a series of large lithographs that in fact illustrated not a single unique "primitive world" but a world that had changed continuously *"in its different periods of formation."* Unger suggested that the lush vegetation in this scene might have been due to an atmosphere that was still somewhat richer in carbon dioxide than in the present world: a hint that the development of the living world might have been linked with that of the Earth itself in some kind of integrated system. The scene encapsulates much of what has been entailed in the discovery of the Earth's deep history ever since the 17th century: exploiting whatever evidence was currently available—in this case, scraps of fossil bones and teeth, plant stems and leaves, and the rocks in which they are preserved—and using it to reconstruct a history that is unavoidably conjectural yet always *corrigible and improvable* in the light of further evidence. In this case the discovery, later in the 19th century, of far more complete fossil remains of the *Iguanodon* showed that the rhino-like horn on the snout had in fact been a claw, and that the whole animal had been substantially different in form from what Unger and his contemporaries imagined.

unexpected and surprising to all concerned. Obvious examples from the more recent part of this narrative were the chance discovery of the Ediacara fossils and those in the Gunflint chert, both of which dramatically changed scientific understanding of the early history of life. The process of historical reasoning about the deep past has produced reconstructions that have always, necessarily, been more or less con-

jectural, simply because they are of events that are unobservable. Yet
they have also always been open to *correction and improvement* in the
light of new evidence: for example, the discovery of fossil specimens
better preserved or more complete than those previously available has
again and again improved the reconstructions, making them more reliable representations of the unobservable deep past.

## HOW RELIABLE IS KNOWLEDGE OF DEEP HISTORY?

This brief history of the discovery of the Earth's own history has now
been brought into the early 21st century. But the past tense—which
is always proper to the work of historians, although many of them
currently seem to think it more sexy to use the present tense—has
been retained right to the end of this narrative, in order to indicate
that the story has certainly not yet reached its conclusion. Many of
the interpretations summarized towards the end of the narrative refer
very much to "work in progress": how, when, and whether the relevant scientists will reach a consensus on these matters, and what it
will be, remain to be seen. There is no reason to treat the present state
of knowledge about the Earth's history as the definitive Truth for all
time, any more than any previous generation of scientists would have
been justified in thinking in the same way about their own ideas. The
phrase "But we now know that . . . ," which is often used by scientists
to dismiss or ridicule the conclusions of their predecessors, is always
liable to return uninvited and embarrass them, a year or a decade or
a century later.

Yet the broad sweep of the historical narrative in this book gives
good grounds for thinking it highly unlikely that any future new discoveries or new ideas will fundamentally undermine or destroy the
major features of the Earth's deep history, as they have gradually been
reconstructed over the past few centuries, though they may well clarify
or modify them quite substantially. In recent decades it has been intel-
lectually fashionable to portray the history of scientific knowledge as
a sequence of radical revolutions and incommensurable "paradigms"
(one of the most overworked terms in current intellectual discourse),
such that what was claimed as sound knowledge in one period might
be almost completely overturned and replaced in a later period. Yet
however apt that model may be in the cases of some other sciences, in

the discovery of the Earth's deep history there has been an unmistakeable trend towards reconstructions and interpretations that have accounted for the available evidence in increasingly satisfactory ways. There has been overall progress. Even the changes of opinion that have often been presented as sudden, dramatic, and "revolutionary"—such as the acceptance of plate tectonics in the 20th century, or the rejection of "catastrophes" in the 19th, or the realization of a lengthy timescale in the 18th—turn out, when studied closely by historians, to have been underlain by much more substantial continuities than those who counted themselves the victors have wanted their contemporaries to believe. As usually happens in the history of the sciences, such controversies have provoked fruitful new lines of research, which have often generated *novel* interpretations incorporating important elements from the "losing" sides as much as from those that claimed to have "won."

One obvious reason for this element of overall progressiveness in the history of a science like geology is the conspicuously cumulative character of the relevant evidence. For example, once a particular fossil specimen has been found, studied, and described, it remains available for future generations to refer back to and incorporate into newer frameworks of interpretation (provided of course that it is not lost or destroyed). One particular fossil shark's tooth that was collected and described by Scilla in the 17th century was incorporated into Woodward's collection (and into his diluvial theory) in the 18th, and re-described and re-interpreted in evolutionary terms by palaeontologists in the 19th and 20th centuries, all in the context of ever-richer stores of comparable specimens. Likewise, fieldwork in regions previously unexplored or at localities previously unexamined has again and again enriched the evidence available to undermine or confirm interpretations old and new. A case in point was the unanticipated discovery that the strange fossils found at Ediacara were not confined to one locality in Australia but characterized a certain distant period of the Earth's history worldwide. Improvements in techniques of investigation have provided, and continue to provide, another source of almost irreversible change that deserves to be called progress. The crucial evidence for the early history of life derived from the Gunflint chert, for example, would have been inaccessible without the earlier development of techniques for studying microscopic features in very

thin sections of any such hard rocks. Above all, the increasingly precise and consistently reliable dating of every part of the Earth's history has depended on ever-improving techniques for analyzing tiny
quantities of radioactive isotopes, notably by harnessing the potential
of the mass spectrometer, an instrument originally devised for quite
different purposes.

It was by no trivial accident of wording that at the end of the 19th
century this new science of *geochronology* derived its name from the
17th-century science of *chronology* (which in fact remains alive and
well wherever events or artifacts from ancient history and ancient cultures are assigned to dates BC or BCE). Both are *historical* projects. In
the 17th century, naturalists such as Steno and Hooke had borrowed
from chronologists such as Scaliger and Ussher: they interpreted the
natural features of rocks and fossils in terms of the Earth's chronology,
which they treated as no different in principle from the chronology
of human history. There was therefore an unbroken intellectual and
conceptual continuity between what the scholarly chronologists were
doing in the 17th century and what geochronologists were continuing to do in the early 21st, despite the obviously huge contrast in the
timescales they were working with. And more than just a continuity:
this deliberate transferring of concepts and methods from culture into
nature—highlighted by the pervasive use of the metaphors of *nature's*
coins and monuments, *nature's* documents and archives—was essential for the development of habits of reasoning that could turn rocks
and fossils, mountains and volcanoes, into the intelligible traces of the
Earth's deep history. Those who deployed these metaphors most effectively and consistently, such as Hooke in the 17th century and Soulavie
in the 18th, were deliberately applying methods and insights borrowed
from the historians of their time. These historians (among whom the
chronologists were simply those who focused particularly on compiling accurately dated historical "annals") were therefore crucially important for the reconstruction of the Earth's own history by those who
were later called geologists.

## GEOLOGY AND GENESIS RE-EVALUATED

The early chronologists have been unjustly maligned in modern times.
This is mainly because they deployed the Bible, and specifically the

book of Genesis, to establish the starting point of their chronologies. But to condemn this as a case of "Religion" distorting or retarding "Science" is to misunderstand what the chronologists were doing. They were trying to plot the world's total history as accurately as possible and from all available historical sources, which in practice meant mainly secular ones. Of course, given their cultural context, most of them interpreted the broad sweep of human history in terms of divine self-disclosure ("revelation") and as a story punctuated by divinely significant "epochs" or decisive moments. But this did not affect the intended character of their chronology *as history*. At its start the chronology was usually derived from the first creation story in Genesis, not only because this provided a dramatic starting point "In the beginning . . . ," but also because the biblical text was believed to be the only historical source that recorded these earliest of all events. Later in the chronology, as relevant secular records became available for less remote times, they were brought in, first as supplementary evidence but later as the main sources for dating events.

A strong case can therefore be made for concluding that the early chronologists' perspective on human history—extending back in the Creation stories, however briefly, to the origin of everything—positively facilitated the later transfer of historical ways of thinking into the study of the natural world. The "hexameral" or six-day narrative of Creation provided a template for the later development of a narrative of the Earth's own history. This template proved its worth when, in the 18th century, it became clear to naturalists that the timescale of the Earth's history must be far longer than the chronologists had assumed. The six "days" were simply expanded into periods of indefinite length—an interpretation sanctioned at the time by biblical scholars— while retaining a sense of the unrepeated directional development of the Earth and its life, all the way from its origin to the present, in an intelligible sequence of events that had culminated in the arrival of human beings. This kind of model made it easy for geologists in the 19th century to treat their science—if they so wished, as many did—as perfectly compatible with their religious practice. On this issue there was no fundamental conflict between the science of geology and the interpretation of biblical texts, at least among those who were aware of current ideas in both fields. In the wider world, those who clung to a "literalistic" approach to the Bible, treating it unhistorically as a set of texts with an unambiguously literal meaning throughout, were relegated to the intellectual and cultural margins, and with good reason.

The myth—for such it is—of a major historical conflict between geology and Genesis is deceptive: the real point of conflict was and is elsewhere. In the 19th century, religious concerns about the implications of the geologists' novel picture of the Earth's vast history were usually underlain by worries about the biologists' equally novel picture of the evolutionary origins of the diversity of living organisms; and these in turn were focused on concerns about the nature and status of the human species in particular. This underlying worry about ourselves was understandable and not altogether misplaced, for the scientific inference about human origins—that human beings have been derived from earlier primates by some kind of purely natural evolutionary process—was increasingly hijacked by those with atheistic agendas to promote. Specifically, Darwin's theory of evolution, which with its idea of natural selection had first provided "transmutation" with a plausible causal explanation, was expanded and transformed by others into the all-embracing "worldview" of *Darwinismus* (the German word captures its pretentious character better than the English!). In the later 19th century, Darwin*ism* or evolution*ism* showed itself to have the potential to become, in effect, an atheistic quasi-religion. This became ever more apparent during the 20th century, when its proponents often displayed an aggressive and dogmatic mind-set that closely matched that of their religious counterparts. Scientists in the early 21st century have often been dismayed by the religious fundamentalists whose political clout, in some parts of the world, was threatening all they stood for. They should have acknowledged their collective failure to puncture the pretensions of those in their own ranks who were peddling an equally pernicious kind of fundamentalism: one that improperly extended the scientific theory of evolution into an atheistic worldview.

To pursue this any further, however, would take this book too far from its proper focus, which has simply been to trace in outline the history of the discovery of the Earth's own deep history. What has been emphasized here has been not so much the vast enlargement of the Earth's timescale, striking though that is, but rather the piecing together and reconstruction of the Earth's unexpectedly eventful history. For the scientists concerned, this has been, and remains, fascinating enough in its own right to absorb their time and attention. For the wider public, what most obviously distinguishes the scientific "ancient Earth" of the 21st century from the traditional "young Earth" of the 17th is not so much the magnitude of the timescale, but rather the place of human beings within the history. Humanity seems to have been reduced from occupying the entire drama (apart from a brief prelude) to being confined to the very last scene.

It would take another book, and a quite different one, to explore fully the cultural impact of this radical change of perspective on what people in the 19th century often called "Man's place in Nature." Here it is sufficient to note that the change was not unprecedented or without an earlier parallel. In earlier centuries, equally dramatic discoveries by astronomers had shifted ideas about the cosmos "From the closed world to the infinite universe" (the title of a classic book on this theme). This had transformed "Man's place in Nature" as radically in the dimension of space as the discoveries summarized in this book later transformed it in the dimension of time. Considered crudely, both reduced humanity in size: the Earth with its human passengers became just one planet orbiting one star in an inconceivably immense space, while human existence on Earth became just the last moment in an inconceivably vast span of time. In neither dimension, however, did these dramatic changes affect the perennial questions surrounding the purpose of human existence and the task of constructing societies, based on both justice and compassion, in which human lives can be lived to the fullest. These existential questions remained much the same in the era of space exploration—whether or not it was believed that "we are alone" in the universe—as they were when it was reasonable to assume that the first human beings were placed in a unique and newly created world only a few thousand years ago. For the many people who continue to tackle these profound questions within a religious context, the vast enlargement of the Earth's history is—*religiously*—no big deal, however impressive and scientifically fascinating they may *and should* find it. Those who choose to live within a well-developed theistic tradition such as Judaism or Christianity (I include myself) can and should take dinosaurs and mass extinctions as much in their stride as they take exoplanets and black holes.

What is certainly untenable is any claim that the discovery of the Earth's deep history has in the past been retarded or obstructed by

"Religion." Of course, in any period of history and in any culture it is possible to find plenty of fools and bigots; but there have also been plenty of those who were neither foolish nor bigoted, both among those who counted themselves religious believers and among those who—often with good reason—criticized the practices of religion in their time. Of course, those for whom a religious perspective gave meaning and purpose to their lives have, in every period, wanted to integrate new scientific knowledge with their existing ideas about the world. But these projects have often provided intellectual templates that have served to extend scientific knowledge rather than constraining it. In every century covered by this narrative, some of those who have contributed most enduringly to the scientific story have also been religious believers. In the history of the discovery of the Earth's own history, as in the history of many other aspects of the sciences, the idea of a perennial and intrinsic "conflict" between "Science" and "Religion"—so essential to the rhetoric of modern fundamentalists, both religious and atheistic—fails to stand up to historical scrutiny.

Where it matters most, however, the new scientific view of the Earth's deep history seems far from having overturned traditional views: those with power in the modern world have singularly failed to comprehend its practical implications. They currently argue, for example, about the direction of climate change in the past decade or two, seemingly unaware of the insignificance of any such short-term trend in the perspective of the far greater changes in the deeper past and likely future. Yet at the same time they seem unaware of the alarming implications of the major mass extinction—a sixth to match the five that have punctuated the past half-billion years—that current policies and practices are causing all around us. This at least is unquestionably "anthropogenic" or man-made. And above all they seem to ignore the impact of the rampant exploitation, over just a few recent decades, of natural resources that accumulated millions or even billions of years ago and are utterly unrenewable. In the light of the history of scientific discovery summarized briefly in this book, such ignorance and disregard for the needs of future human generations are surely inexcusable.

To conclude, however, on a more positive note: during the past three or four centuries, the imaginative but also scrupulously careful work of those who have called themselves savants, naturalists, or scientists—many of whom, to repeat the point, have been devoutly religious people—has transformed our view of our human place in the natural world, by reconstructing the amazingly eventful deep history of the Earth and its life with ever more robust and reliable evidence. This must surely count as one of the most impressive scientific achievements of all time. I hope this brief account of its history will help to make it more widely understood and appreciated. Certainly it deserves to be.

APPENDIX

# Creationists out of Their Depth

This book has traced the *history* of the gradual discovery of the
Earth's own deep history; it has not presumed to offer a summary of current scientific knowledge. But there is one strange
feature of the current scene that demands historical comment:
so strange, and so far outside the mainstream of scientific thinking and practice, that it is appropriate to describe it in an appendix. This is the appearance in recent decades in the United States
(and only derivatively elsewhere in the world) of the movement
known as *"creationism,"* which rejects almost all aspects of the
interpretation of the history of the Earth and its life that has
been worked out over the past three or four centuries by those
who are now called scientists. What is most prominent in creationism is its vehement rejection of evolution, and particularly
of what are alleged to be the implications of evolutionary theory
for understanding human beings. But what is also prominent is
its startling re-invention of the idea of a "young Earth," which
the sciences of the Earth outgrew for very good reasons back in
the 18th century. The following brief summary of the history of
creationism places its bizarre rejection of the mainstream Earth
sciences in the context of its even more striking clashes with the
life sciences and particularly with the idea of evolution.

The first chapters of this book, describing the early development of ideas about the history of the Earth and its life, showed how 17th-century chronologists used the Bible as one of their sources for constructing a timeline of world history. Probing back from ancient Rome and Greece into even earlier times, they believed that biblical texts eventually became the only historical records available. In particular, the two Creation stories in Genesis were taken to be the only accounts of the earliest times of all; they were thought to be reports of what must, necessarily, have been disclosed by God directly to Adam or his descendants. But although the rest of the Bible was also believed to be in some sense divinely "inspired," it was recognized to be a collection of diverse texts written or recorded *by human beings*: in the case of Genesis, reputedly by Moses (likewise, in my own mainstream religious tradition, rooted in the Protestant Reformation, the 16th-century liturgy encouraged worshippers to "Hear what Saint John saith . . . ," for example, *not* to "Hear what the Bible saith . . ."). Back in the early or "Patristic" centuries AD/CE the variety of possible levels of interpretation of specific biblical texts had already been well explored; the "literal" was just one of several, and not the most highly valued. Furthermore, the principle of "accommodation" had acknowledged that the language used in the Bible, even if divinely inspired, must necessarily have been adapted to the capacities of the original audience of any given text, or its meaning or message would not have been understood. In much later centuries the deepening of biblical scholarship, and a growing historical awareness of the "otherness" of earlier cultures, led scholars and theologians to recognize, for example, that the "days" of the first Creation story might not have referred to days in the modern sense, and that the recorded universality of Noah's Flood might have referred to the world as it was known at first hand to the story's original audience. Above all, it was acknowledged and indeed emphasized that the primary purpose of the Bible was to record and interpret the historical events on which central Christian concepts such as Incarnation and Redemption were founded, and to point to their practical implications for everyday living; it was not to instruct humanity in any of the sciences. As Galileo was said to have quipped, the Bible was there to show us how to go to Heaven, not how the heavens go.

In the light of this long historical tradition of scholarly "hermeneutics" or methods of interpretation, the revival of biblical "literalism" in the late 19th and early 20th centuries, most prominently in American Protestantism, was surprising to the rest of the Christian world. (Its simultaneous and equally strange parallel in worldwide Catholicism was the revival of cults based on local miracles, as at Lourdes, likewise paradoxically characteristic of a newly scientific and technological age.) More particularly the idea of the absolute verbal "inerrancy" of the Bible, as propounded by some American religious figures, was a startling innovation. However, this new literalism was an understandable and to some extent justifiable reaction to ultra-"liberal" movements elsewhere in American religious life, which had abandoned any element of transcendence and reduced Christianity to little more than what even its proponents called a merely "social gospel." Early in the 20th century the famous series of booklets entitled *The Fundamentals* (1910–15), which later gave its name to *"fundamentalism,"* was designed to resist this trend by restating basic Christian doctrines; the primary target was ultra-liberal theology and the reductive variety of biblical criticism that underlay it, not scientific ideas as such. But after the First World War (which the United States had entered belatedly though with decisive effect), the politician William Jennings Bryan led a moral crusade in which the horrific brutality of the war itself and all the social ills of postwar modernity were blamed on the supposedly atheistic implications of the idea of evolution when it was extended to human beings.

This was the background to a high-profile trial in Tennessee in 1925, in which Bryan—a moralist rather than a biblical literalist—led the successful prosecution of John Scopes for having broken state law by teaching human evolution in his biology classes. Although Scopes's defense lawyer, Clarence Darrow, was widely considered to have won a moral victory, Bryan's stance became the inspiration for a growing fundamentalist movement in American Protestantism in the following decades (Bryan himself died soon after the Scopes trial). A combination of diverse factors specific to American social and political life lay behind this trend, which had no close parallel in the rest of the world: resentment of South against North, long-established Protestants against immigrant Catholics, conservative rural societies against sophisticated urban culture, the less well educated against academic elites, and so on. Above all there was the distinctively American constitutional separation of church and state, which was crucial to the question of what could or should be taught in the public education system.

The idea of evolution became the fundamentalists' main scientific target; and particularly what was understood as Darwinism, owing to its often reductive application to the origin and nature of human beings (that in all respects we are *"nothing but"* naked apes). The Earth's extremely lengthy history, and the fossil record that scientists interpreted as evidence for long-term and large-scale evolution, were recognized as essential to the scientific case that the fundamentalists needed to undermine. At the turn of the century the Adventist writer George McCready Price had found inspiration in the ideas of the American founder of his sect, to argue that the basic principles of geology—of which he had minimal practical experience—were fatally flawed. He claimed that the science's entire case for the long history of the Earth and its life could be recast in terms of a six-day Creation only a few thousand years ago, followed by a very brief worldwide Flood in which the whole pile of rock formations had been deposited all at once; the latter had a startling similarity to Woodward's diluvial theory two centuries earlier. Among Price's books expounding this revived "young Earth" view, *The New Geology* (1923) had the greatest impact on the Protestant religious public in America, though the leading geologist Charles Schuchert epitomized scientific opinion about it when he dismissed Price as "harboring a geological nightmare." Undeterred, Price and his friends continued their crusade for a "young Earth" that would make any kind of evolution almost impossible for sheer lack of time. In the 1940s they formed a Deluge Geology Society in California, but they were criticized from within the American Scientific Affiliation—a body set up in 1941 to represent scientists who were theologically conservative Christians—for failing to take account of the strong geological evidence for an "ancient Earth."

The future of young-Earth creationism therefore looked unpromising, and remained largely confined to Adventists, until the publication of *The Genesis Flood* (1961) by the Bible teacher John Whitcomb and the engineer Henry Morris: both came from fundamentalist backgrounds and had no greater experience of geology than Price had had before them. The book's unanticipated success, and its impact on a wider range of Protestant opinion in America, led in 1963 to the foundation of the Creation Research Society; its membership was restricted to those with scientific qualifications—though not necessarily in the sciences relevant to the contentious issues—and it kept itself doctrinally on the straight and narrow path of biblical inerrancy. In the 1970s, however, a divergence in tactics became apparent within the creationist movement. Some creationists continued to focus on finding evidence for a very recent Creation and a worldwide Flood. For example, they sought to undermine "ancient Earth" stratigraphy and palaeontology by identifying Mesozoic dinosaur footprints as those of early human beings, and by interpreting the Grand Canyon as the product of ultra-rapid deposition of the huge pile of rock formations followed immediately by ultra-rapid erosion of the deep canyon through them as the water drained away; and they searched diligently for the remains of Noah's Ark high on the slopes of Ararat. Others, in a major tactical shift, campaigned for creationism to be given "equal time" with "evolutionism" in American public education, on the grounds that they were alternative theories that were scientifically equivalent and equally entitled to a hearing. Whereas the first group continued to argue that all the evidence of geology and palaeontology could be reinterpreted in line with a narrowly literal reading of the Genesis narratives, the second downplayed Genesis altogether and sought to rebrand itself as strictly scientific. Revealingly, Morris's schools textbook *Scientific Creationism* (1974) was published in two versions: one, for public schools, with no reference whatever to the Bible; the other, for "Christian" (i.e., fundamentalist) schools, with an extra chapter on "Creation according to scripture." And "scientific creationism" later rebranded itself as "creation science": to its critics, a glaring oxymoron.

The subsequent history of creationism, into the early 21st century, was marked by a sequence of very public court cases—which made headline news in the United States though they were little noticed elsewhere—in which creationists' claims to be given "equal time" in American public education were aired repeatedly. A further tactical shift appeared in the 1990s with the emergence of a fresh set of creationist arguments. *"Intelligent Design,"* which was given some scientific appearance by the biochemist Michael Behe in *Darwin's Black Box* (1996), simply rewarmed the traditional "argument from design." It extended this specific variant of natural theology, as formulated for example by Paley at the start of the 19th century, from the level of whole organisms and their constituent organs down to that of micro-structures and molecular mechanisms within the living cell. But biol-ogists swiftly pointed out that the allegedly "irreduceable complexity" of these features was just as much open to interpretation in evolution-ary terms as, for example, the amazingly complex adaptations of the human eye had proved to be in the 19th and 20th centuries. Nonethe-less, Intelligent Design gave creationism a tactical boost in the early 21st century. It prudently concealed the movement's roots in bibli-cal literalism; it enhanced—at least in the eyes of the non-scientific public—its claim to be a legitimate "science"; and it played down its insecure reliance on "young Earth" geology. Arguments for an Earth created from scratch only a few millennia ago, and one ravaged by a worldwide and ultra-catastrophic Flood still more recently, had had to rely ever more obviously on assuming a radical disjunction between the present world and all its earlier history, involving major changes in some of the most basic physical "laws of nature." No one had been so rash, at least not since Woodward in the 17th century invoked a temporary suspension of Newton's universal gravitation to help ac-count for his worldwide Deluge: that was the scale of what, implausi-bly, "young Earth" creationism required.

As emphasized repeatedly here, creationism in its manifold varie-ties has been a movement as all-American as motherhood and apple pie; scientists elsewhere in the world were often astonished and even incredulous when told by their American colleagues about the lat-est activities of creationists in the United States. Only in the late 20th century was creationism exported to other parts of the world, usually with massive financial support from American fundamentalists. In contrast, indigenous creationist movements in other countries such as Britain were mostly small in scale, limited in impact and brief in duration, unless and until they received such external support. Only in the early 21st century did creationism begin to take root on a larger scale and even spread beyond Christian fundamentalism to analogous fundamentalist movements in Judaism, Islam, and other religious tra-ditions. What all these movements have strikingly in common is a tight linkage between their rejection of the very idea of evolution and their vehement hostility to a characteristic cluster of other supposed evils of modernity such as divorce, abortion, homosexuality, and even feminism. Creationism has come to be linked, most obviously in the United States, with a specific kind of political ideology.

In a wider context the persistence of "young Earth" ideas is now most closely analogous to the persistence (among a far smaller minority of the public) of a belief that in reality the Earth is flat, and not at all a ball in space. Young-Earthers are now unmistakeably equivalent—philosophically—to flat-Earthers, and proponents of Intelligent Design are equally out of touch. For all the noise that creationism generates, it is no more than a bizarre sideshow that has set itself in implacable opposition to one of the most solid and reliable of human scientific achievements. Sadly, creationists are utterly out of their depth.

# GLOSSARY

Words printed in *italics* have their own entries in this glossary.

**Actualism**   The method of reasoning that treats "actual causes," i.e. processes observably active at the present day, as the most reliable key for interpreting the traces of the deep past.

**Aeon**   The longest conventional unit of geological time; only the most recent of four aeons, the *Phanerozoic*, has a fairly rich *fossil record*.

**Alluvial**   Those deposits at the surface, overlying rock *formations* of all kinds and clearly composed of debris derived from them, and therefore of more recent origin.

**Ammonites**   Distinctive *fossil* mollusc shells, usually coiled in a plane spiral, highly diverse in form and often abundant in *Mesozoic formations*, but totally extinct.

**Anthropocene**   Recently proposed as the very brief and still current *epoch* in the Earth's history, during which the physical impact of the human species has become a major factor.

**Antiquary**   A scholar or *savant* who made a special study of *antiquities*; many would now be described as archaeologists.

**Antiquities**   Human artifacts surviving from the past, and particularly from the ancient worlds of Greece and Rome or even earlier periods of human history.

**Archaean**   The earliest Precambrian *aeon* from which substantial rocks survive, most of them *metamorphic*; formerly classed as *Primary* rocks.

**Argument from design**    One of the traditional philosophical grounds for the reality of God, based on the apparently "designed" character of the world, and especially of organisms.

**Astrobleme**    The trace of an impact event on the Earth's surface, either a well-preserved crater or a circular rock structure interpreted as the eroded remains of a crater.

**Astrogeology**    The application, to the Moon and other bodies in the Solar System (and even beyond it), of methods and concepts developed by the science of geology on the Earth.

**Atlantis**    A legendary inhabited land, conjecturally located either within the Mediterranean or in the Atlantic, which had subsequently been submerged beneath the sea.

**Basalt**    A hard, dark and fine-grained rock, formerly of uncertain origin, and claimed by *vulcanists* as a volcanic lava, but by *neptunists* as a hardened sediment.

**Belemnites**    Distinctive *fossils* of bullet-like shape, forming a part of chambered mollusc shells related to *ammonites*; characteristic of *Mesozoic formations* and wholly extinct.

**Cambrian explosion**    The relatively sudden appearance of a diversity of quite large animals around the start of the *Cambrian period* and at the very end of the preceding *Precambrian*.

**Cambrian period**    The first and earliest *period* in the *Palaeozoic era*, with *fossils* representing a diversity of quite large animals, in contrast to the preceding *Precambrian*.

**Catastrophism**    The theory that assigns an important role in the Earth's history to occasional sudden events of great intensity, some more intense than any witnessed in human history.

**Cenozoic**    The most recent of the three *eras* within the *Phanerozoic aeon* of the Earth's history; it includes, for example, the time in which mammals have been abundant.

**Chronology**    The branch of historical work that uses sources of all kinds to fix precise dates for past events in human history, which may then be tabulated as annals or year-by-year records.

**Continental shelves**    Areas that are geologically parts of continents, but are covered—often, in geological terms, only intermittently—by relatively shallow seas.

**Crust**    The uppermost or outermost major layer of the Earth as a solid body; the "basement" (or formerly *Primary*) rocks of which both continents and ocean floors are composed.

**Deism**    The kind of theology that conceives of God as having created the universe but then left it to run on its own according to timeless "laws of nature."

**Deluge, geological**    The event that was widely believed to have devastated parts of the Earth in the geologically recent past, often conceived as some kind of large-scale tsunami.

**Diluvial**   Related to the *"Deluge"* that was thought to have affected the whole Earth or substantial parts of it; it might or might not be identified with the *Flood* recorded in *Genesis*.

**Diluvium**   The *Superficial* deposits attributed to the geological *Deluge* (not necessarily to the biblical *Flood*); later reinterpreted as the traces of former glaciers and ice sheets.

**Epoch**   Originally, a decisive dated turning-point in human history; later, in geology, a conventional unit of the Earth's history, as a subdivision of a *period*.

**Era**   The second largest conventional unit of geological time, for example the *Palaeozoic*, *Mesozoic*, and *Cenozoic eras* that together comprise the *Phanerozoic aeon*.

**Erratic blocks**   Large blocks of rock that have evidently been shifted, perhaps by *"diluvial"* or glacial processes, from distant areas where the same kinds of rock form the bedrock.

**Eternalism**   The philosophical claim that the universe—and the Earth as a part of it—have existed and will exist from and to eternity, without any initial act of creation or final end.

**Euhemerist**   The historical method that interprets ancient myths of gods and superhuman heroes as garbled records of memorable human beings or natural events.

**Exoplanets**   Planets orbiting other stars far beyond the Solar System; they may be "gas giants" like Jupiter or rocky and Earth-like.

**Fixism**   The theory that continents have not changed in position relative to each other in the course of the Earth's history; in contrast, *mobilism* claims that they have.

**Flood, biblical**   The event recorded in the book of *Genesis*, which might or might not be identified as the same event as the "geological *Deluge*."

**Formation**   A set of rocks of distinctive character, the outcrop of which could be traced on the ground, and its extension beneath the surface confirmed by wells and boreholes.

**Fossil record**   The totality of the traces of organisms and their activities preserved as *fossils* in successive rock *formations*, recording—however fragmentarily—the history of life.

**Fossils**   Originally, just "things dug up" (as in modern "fossil fuels"); later, specifically those of organic origin, the remains or traces of living organisms preserved in rocks.

**Genesis**   The first of the five books of the Pentateuch, the core of the Jewish scriptures and the first part of the Christian "Old Testament," traditionally attributed to Moses.

**Geochronology**   The chronology of the Earth's history, established by dating its events (now usually by *radiometric* methods); analogous to the *chronology* of human history.

**Geognosy**   The science that focused on the three-dimensional structure of the rocks visible at the Earth's surface and (in mines) beneath it; practiced by "geognosts."

**Geophysics**   The science that focuses on the physical properties and processes of the Earth, particularly those in its unseen interior, detectible mainly by instrumental methods.

**Geothermal gradient**   The rate at which the temperature of rocks rises with depth, e.g. in mines, indicating the Earth's "internal heat" (whatever the source of the heat).

**Glossopetrae**   The tongue-shaped stony objects that, being closely similar to sharks' teeth, were persuasive evidence that at least some *fossils* had been parts of living organisms.

**Great oxygenation event**   The time at which, early in the *Proterozoic aeon*, free oxygen began to accumulate in the Earth's atmosphere.

**Hadean**   The conjectural *Precambrian aeon* at the start of the planet's history, during which a "heavy bombardment" of major asteroid or cometary impacts was intense, as on the Moon.

**Hexameron**   The "six-day" Creation narrative in *Genesis*, followed and completed by a seventh or sabbath day of God's rest.

**Interglacial**   A period of relatively warm climate during an Ice Age, sandwiched between glacial periods of colder climate when ice sheets extended widely.

**Isostasy**   The *geophysical* theory according to which the lighter *crustal* rocks of continents are "in balance," in effect floating or buoyant on deeper and denser layers.

**Julian timescale**   An artificial timescale devised in the 16th century to provide an ideologically neutral timeline on which all the events of world history could be plotted.

**K/T boundary**   The boundary between the Cretaceous *period* ("Kreide" in German) and *Tertiary* time, marking an apparent mass extinction event.

**Land-bridges**   Possible former connections between continents, like the present Isthmus of Panama, by which terrestrial animals and plants could have spread across oceans.

**Living fossils**   Organisms that have been discovered alive, after their relatives were well known as *fossils* and were assumed to have been long extinct.

**Ma**   An abbreviation for "millions of years" (usually meaning "before the present") as estimated by various methods of dating, now usually *radiometric*.

**Macro-evolution**   Evolutionary change linking different major groups of animals or plants, e.g. reptiles and birds; in contrast, the origin of new species involves "micro-evolution."

**Man of science**   The gendered (but at the time factually accurate) term widely used in the 19th century to denote those who were later called "scientists."

**Mantle**   The region of the Earth's deep interior that underlies the *crust* of both continents and oceans; it is composed of solid rock, in contrast to an innermost liquid "core."

**Mass spectrometer**   An instrument that allows extremely accurate chemical analyses; it is used to measure radioactive isotopes in minerals, and thereby to derive *radiometric* dates.

**Megafauna**   A set of unusually large animals, particularly the large mammals (mammoths, mastodons, etc.) characteristic of the *Pleistocene* Ice Age.

**Mesozoic**   The middle of the three *eras* within the *Phanerozoic aeon* of the Earth's history; it includes, for example, the time in which dinosaurs and other diverse reptiles were abundant.

**Metamorphic rocks**   Those that have been radically altered in mineral composition by being subjected to intense heat and pressure in the depths of the Earth.

**Mobilism**   The theory that continents and *crustal* "tectonic plates" have moved in position relative to each other in the course of the Earth's history; in contrast, "*fixism*" denied such movements.

**Moraines**   Ridges composed of boulders on the edges of glaciers and ice sheets, derived from material plucked from the bedrock, transported within the ice, and dropped where it melts.

**Nappe**   A huge mass of rocks that has been thrust over others, like a crumpled tablecloth (in French, a "nappe"), during *orogenic* movements of the Earth's *crust*.

**Natural chronometer**   Any method that allows geological time to be measured (in years or any other units), on evidence drawn from the uniform rate of some natural process.

**Naturalist**   The kind of *savant* who studied "natural history," or the diversity of things "animal, vegetable, and mineral" (i.e., in later terms, zoology, botany, and geology).

**Natural theology**   The branch of theology that analyzes claims about the relation between God and the natural world, including human nature.

**Natural philosopher**   The kind of *savant* who studied natural philosophy, or the causes of natural phenomena of all kinds, often referred to just as "philosophy."

**Neptunists**   *Naturalists* who interpreted certain distinctive rocks, notably *basalt*, as being hardened sediments and not (as *Vulcanists* claimed) volcanic products.

**Orogeny**   A time of major movements in the Earth's *crust* along some linear zone, resulting in the elevation of a new range of mountains.

**Palaeo-magnetism**   The magnetic field in mineral crystals within rocks, recording the orientation of the Earth's magnetic field at the time and place where the rock was formed.

**Palaeozoic**   The earliest of the three *eras* comprising the *Phanerozoic aeon* in the Earth's history; it includes, for example, the time in which *trilobites* were abundant.

**Period**   A conventional division of geological time within an *era*, for example the *Cambrian* and "Silurian" *periods* within the *Palaeozoic*.

**Phanerozoic**   The most recent *aeon* in the Earth's history, comprising all the time since the end of the *Precambrian*, or the totality of the *Palaeozoic*, *Mesozoic*, and *Cenozoic eras*.

**Plate tectonics**   The *mobilist* theory according to which the Earth's *crust* is composed of separate "plates," the movements of which have shifted continents in relative position.

**Pleistocene**   The recent *epoch* within the *Cenozoic era* of the Earth's history, comprising the most recent major "Ice Age" or sequence of glacial and *interglacial* episodes.

**Plurality of worlds**   The speculation that the Earth may not be the only body in the universe that supports human-like intelligent life; its modern manifestation is the *SETI* program.

**Precambrian**   The totality of all the Earth's history before the start of the *Cambrian period* and of the *Phanerozoic aeon*; it comprises the *Proterozoic*, *Archaean*, and *Hadean* aeons.

**Prehistory**   The totality of human history before the rise of literate civilizations: divided in the 19th century into the "Stone Age," "Bronze Age," and "Iron Age."

**Primary**   Those rocks found deepest in the Earth (though rising to the surface in some areas), without *fossils*; they were inferred to have been formed earliest in the Earth's history.

**Proterozoic**   The youngest *Precambrian aeon*, during which the *great oxygenation event* marked the start of an atmosphere with oxygen; with a sparse record of microscopic life.

**Quaternary**   The most recent part of the *Cenozoic era*, subsequent to the *Tertiary*; it comprises the *Pleistocene epoch*, the post-glacial "Holocene," and now the *Anthropocene*.

**Radiometric**   The method of estimating the dates, in years, of rocks and minerals, and hence of events in the Earth's history, based on the constant rates of decay of radioactive isotopes.

**Raised beaches**   Former beaches now high above present beaches, showing geologically recent uplift of the landmass or lowering of the sea level.

**Savant**   An educated intellectual, knowledgeable in any or all of (in modern terms) the natural and human sciences and humanities.

**Secondary**   Those rock formations found overlying, and sometimes composed of debris of, *Primary* rocks and therefore inferred to be of later origin; often containing *fossils*.

**Sedimentary**   Those rocks that have been formed, under water or on land, by the deposition of mineral or organic material, usually in successive layers or *strata*.

**SETI**   The "Search for Extra-Terrestrial Intelligence" elsewhere in the universe beyond the Earth: not just life, but highly complex life presumed to be technologically advanced.

**Shocked quartz**   A variety of quartz (also called "coesite") formed only under conditions of extremely high pressure, as in a nuclear explosion or in an impact from space.

**Snowball Earth** An informal term to denote the Earth during those episodes when the planet is thought to have been largely or even completely covered with snow or ice sheets.

**Strata** The layers of rock, originally formed as successive layers of sediment, which usually comprise a *formation* of "stratified" sedimentary rock such as sandstone, shale, or limestone.

**Stratigraphy** The science that describes and compares sequences of rock *formations* in different regions; similar to the former science of *geognosy*.

**Subduction** The process of *plate tectonics* by which the material of a *crustal* plate is dragged downwards by a descending convection current in the underlying *mantle* rocks.

**Superficial deposits** Those at the Earth's surface, including *alluvial deposits* but also earlier deposits that might be distinguished as *diluvial* and be attributed to a geological "*Deluge.*"

**Tertiary** Those rock formations previously classed as *Secondary*, but overlying the distinctive Chalk *formation* (yet underlying any *Superficial* deposits); later assigned to the *Cenozoic era*.

**Theism** A theology that conceives of God as transcendent and as creating the universe, but also holding it continuously in existence and interacting with it throughout.

**Theory of the Earth** The specific kind of theory that aspired to explain how the Earth works as a total "system," in past, present, and future, according to the timeless "laws of nature."

**Till** A distinctive *Superficial* deposit, also known as "boulder clay," in which angular blocks of rock of all kinds and sizes are embedded; if the clay is solidified, it is a "tillite."

**Transition** Those rock *formations* transitional in position between *Primaries* and *Secondaries*, containing only a few *fossils*; they were therefore inferred to be also intermediate in age.

**Trilobites** Animals with jointed external skeletons, diverse in size and form, often abundant as *fossils* in *Palaeozoic* strata, but totally extinct by the end of that era.

**Uniformitarianism** The theory that the Earth has been in a "steady state" throughout its deep history, with events and processes uniformly like those of the present, in both rate and intensity.

**Varves** Distinctive layers of sediment deposited annually in lakes at the edges of glaciers and ice sheets; used to construct an accurate *geochronology* of recent Earth history.

**Vulcanists** *Naturalists* who interpreted certain distinctive rocks, notably *basalt*, as being volcanic products and not (as *Neptunists* claimed) hardened sediments.

# FURTHER READING

This book has been written primarily for those who are unlikely to have the time to pursue its theme any further, but others may value some suggestions for reading more on particular topics. The publications mentioned below have been chosen as some that are based on scholarly research but are relatively accessible to readers without background knowledge of either the history or the science involved. Some of these works have come to be regarded as "classics" in their field; others are more recent, and their notes and bibliographies give detailed references to the modern research on which they (and the present book) are based. The publications listed below are limited to those written in English or available in English translation; and they are mostly books, rather than the scholarly articles—often less accessible—in which some of the most important research in this field is published. Although some items cover large parts of the history of the Earth sciences, this selection has been chosen for relevance to the specific theme of this book, namely the history of the discovery and reconstruction of the Earth's own history. And books dealing with very recent research are restricted to those that are explicitly historical in approach, since there are plenty of excellent books describing *current* scientific knowledge of the Earth's history. Fuller details of all the items mentioned are given in the bibliography; many are now available in digital formats.

## Works spanning the whole period (17th to early 21st centuries)

Richet, *Natural History of Time*, and Wyse Jackson, *Chronologers' Quest*, are reliable histories of ideas about the Earth's timescale; Gorst, *Aeons*, also links these ideas with the cosmological timescale of the universe; all three books cover the whole of human history from Antiquity to the 20th century. Lewis and Knell, *The Age of the Earth*, includes many useful articles on the history of the debates, from the 17th through the 20th centuries.

Gould, *Time's Arrow, Time's Cycle*, includes insightful analyses of, particularly, the "theories of the Earth" of Burnet, Hutton, and Lyell. Scattered through Gould's several well-known volumes of short essays on natural history are many valuable pieces on particular figures in the history of the Earth sciences. Huggett, *Cataclysms and Earth History*, analyzes the many varieties of "diluvial" theory from Antiquity to the 20th-century revival of catastrophism.

Kölbl-Ebert, *Geology and Religion*, includes, among a variety of articles, K. V. Magruder on "The [17th-century] idiom of a six-day creation"; Martin Rudwick on "Biblical flood and [19th-century] geological deluge"; also R. A. Peters on "Theodicic creationism," a former creationist's notable attempt to explicate creationism as a project deeper than naïve literalism.

*Earth Sciences History* is an international journal publishing scholarly articles, many of which are relevant to the theme of this book.

### Early period (from the 17th to the middle of the 18th century)

Rossi, *Dark Abyss of Time*, is a classic account, fully international, ranging from Hooke to early Buffon. Porter, *Making of Geology*, is another classic, focused on Britain, and extending to the time of Hutton. Rappaport, *When Geologists Were Historians*, is an outstanding study of this period, focused on the Francophone world. Poole, *World Makers*, is a more recent cultural history of the 17th-century English savants who theorized about the Earth.

Grafton, *Defenders of the Text*, is an authoritative review of the intellectual world in which Scaliger's science of chronology was embedded. Impey and MacGregor, *Origins of Museums*, is a classic collection of articles on "cabinets of curiosities." Rudwick, *Meaning of Fossils*, chapters 1 and 2, describes early debates on the nature of "fossils" and their interpretation as natural antiquities. Cutler, *Seashell on the Mountaintop*, is a biography of Steno, popular in style but reliable in content.

### Middle period (from the mid-18th to the later 19th century)

The middle chapters of this book are in effect a greatly condensed version of Rudwick, *Bursting the Limits of Time* and its sequel *Worlds Before Adam*, which jointly give a detailed but accessible narrative with abundant illustrations from the original sources; Rudwick, *New Science of Geology* and *Lyell and Darwin*, reprint many articles on specific issues.

Roger, *Buffon*, is a fine biography by the leading Buffon scholar of the 20th century. Heilbron and Sigrist, *Jean-André Deluc*, reviews recent research on Deluc. Corsi, *Age of Lamarck*, is a classic account of evolutionary theorizing before Darwin. Rudwick, *Georges Cuvier*, translates and comments on Cuvier's most important work on fossils. James Secord's abridged edition of Lyell's *Principles of Geology* includes a valuable introduction to this major work and its social context. Herbert, *Charles Darwin, Geologist*, is a detailed study of Darwin's first scientific career.

Rudwick, *Meaning of Fossils*, chapters 3 through 5, describes the use of fossils as traces of deep history and of life's evolution, to the mid-19th century; Rudwick, *Scenes from Deep Time*, reproduces many early examples of pictorial reconstructions, with commentaries on their iconography. Grayson, *Establishment of Human Antiquity*, is a classic account of the 19th-century debates; Van Riper, *Men among the Mammoths*, is particularly focused on the decisive mid-century British research.

O'Connor, *Earth on Show*, is the finest (and most entertaining) cultural history of the relations between geologists in Britain and those who used their work in the wider public realm of "popular" science, including literalist or "scriptural" writers. Jordanova and Porter, *Images of the Earth*, includes, among other useful articles, John Hedley Brooke on "The natural theology of the geologists," and David Allen on geology's relation to other natural history sciences. Rudwick, *Great Devonian Controversy*, is a detailed narrative that exemplifies the character of "expert" geological debate at this period.

Imbrie and Imbrie, *Ice Ages*, traces the 19th-century recognition of the historical reality of the Pleistocene glaciations, and 20th-century debates on their causes. Greene, *Geology in the Nineteenth Century*, describes theories of mountain-building and global tectonics in the era before Wegener. Burchfield, *Lord Kelvin and the Age of the Earth*, is a classic account of the arguments before the discovery of radioactivity. Bowler, *Theories of Human Evolution*, covers early 20th-century theorizing and its roots in the 19th.

### Later period (from the late 19th to the early 21st century)

Bowler and Pickstone, *Modern Biological and Earth Sciences*, includes useful summaries by Mott Greene on "Geology," Ronald Rainger on "Paleontology," and Henry Frankel on "Plate tectonics." Oldroyd, *The Earth Inside and Out*, includes Cherry Lewis on "Arthur Holmes' unifying theory" linking radiometric dating to continental drift, and Ursula Marvin on "Geology: from an Earth to a planetary science." Krige and Pestre, *Science in the Twentieth Century*, includes R. E. Doel on "The earth sciences and geophysics," which is particularly useful on the new planetary perspective.

Lewis, *Dating Game*, is a biography of Arthur Holmes and an account of his work on radiometric dating. Hallam, *Revolution in the Earth Sciences*, is a geologist's fine history of both continental drift and plate tectonics, written soon after the dust had settled. LeGrand, *Drifting Continents and Shifting Theories*, is an account that evaluates the issues in terms of the growth of scientific knowledge. Oreskes, *Rejection of Continental Drift*, is an outstanding history and analysis of the theory, by a historian originally trained in the sciences, focusing on the theory's initially negative reception by US scientists, but also describing its development outside North America; her *Plate Tectonics*, with her own introduction, is an important collection of essays by many of the leading participants.

Schopf, *Cradle of Life*, includes (chapters 1 and 2) a participant's valuable history of the discovery and interpretation of Precambrian fossils; Brasier, *Darwin's Lost World*, is another participant's more informal account. Arnaud et al., *Neoproterozoic Glaciations*, includes "A history of Neoproterozoic glacial geology, 1871–1997" by Paul F. Hoffman, a major participant in the recognition of Precambrian ice ages.

Baker, "Channeled Scabland," is a historical account of this early "neocatastrophist" controversy. Sepkoski, *Rereading the Fossil Record*, is a fine history of the quantitative "paleobiology" movement that identified likely mass extinction events; Sepkoski and Ruse, *Paleobiological Revolution*, includes, among a collection of valuable essays by leading palaeontologists, Susan Turner and David Oldroyd on "Reg Sprigg and the discovery of the Ediacara fauna." Glen, *Mass Extinction Debates*, is a similar collection of essays, with the editor's interpretation of the argument. Raup, *Nemesis Affair*, is a lively participant's account of the then ongoing argument about major impacts from space.

Numbers, *Creationists*, is the standard history of the movement, revised to cover recent "Intelligent Design" arguments. Marty and Appleby, *Fundamentalisms and Society*, includes James Moore on "The creationist cosmos of Protestant fundamentalism," a valuable interpretation of its historical roots. Schneiderman and Allmon, *For the Rock Record*, on geologists' responses to "young-Earth" and "intelligent design" arguments, includes Timothy H. Heaton's historical review of "Creationist perspectives on geology."

# BIBLIOGRAPHY

Arnaud, Emmanuele, Galen P. Halverson, and Graham Shields-Zhou (eds.), *The Geological Record of Neoproterozoic Glaciations*, Geological Society, 2011.

Baker, Victor R., "The Channeled Scabland: a retrospective," *Annual Reviews of Earth and Planetary Sciences*, vol. 37, pp. 393–411, 2009.

Bowler, Peter J., *Theories of Human Evolution: A Century of Debate, 1844–1944*, Basil Blackwell, 1986.

Bowler, Peter J., and John V. Pickstone (eds.), *The Modern Biological and Earth Sciences* [Cambridge History of Science, vol. 6], Cambridge University Press, 2009.

Brasier, Martin, *Darwin's Lost World: The Hidden History of Animal Life*, Oxford University Press, 2009.

Burchfield, Joe D., *Lord Kelvin and the Age of the Earth*, Science History, 1975.

Corsi, Pietro, *The Age of Lamarck: Evolutionary Theories in France, 1790–1830*, University of California Press, 1988 [*Oltre il Mito*, Il Mulino, 1983].

Cutler, Alan, *The Seashell on the Mountaintop: A Story of Science, Sainthood, and the Humble Genius Who Discovered a New History of the Earth*, Dutton, 2003.

Glen, William, *The Mass Extinction Debates: How Science Works in a Crisis*, Stanford University Press, 1994.

Gorst, Martin, *Aeons: The Search for the Beginning of Time*, Fourth Estate, 2001.

Gould, Stephen Jay, *Time's Arrow, Time's Cycle: Myth and Metaphor in the Discovery of Geological Time*, Harvard University Press, 1987.

Grafton, Anthony T., *Defenders of the Text: The Traditions of Scholarship in an Age of Science, 1450–1800*, Harvard University Press, 1991.

Grayson, Donald K., *The Establishment of Human Antiquity*, Academic Press, 1983.

Greene, Mott. T., *Geology in the Nineteenth Century: Changing Views of a Changing World*, Cornell University Press, 1982.

Hallam, A., *A Revolution in the Earth Sciences: From Continental Drift to Plate Tectonics*, Clarendon Press, 1973.

Heilbron, J. L., and René Sigrist (eds.), *Jean-André Deluc: Historian of Earth and Man*, Slatkine Érudition, 2011.

Herbert, Sandra, *Charles Darwin, Geologist*, Cornell University Press, 2005.

Huggett, Richard, *Cataclysms and Earth History: The Development of Diluvialism*, Clarendon Press, 1989.

Imbrie, John, and Katherine Palmer Imbrie, *Ice Ages: Solving the Mystery*, Harvard University Press, 1986.

Impey, Oliver, and Arthur MacGregor (eds.), *The Origins of Museums: The Cabinet of Curiosities in Sixteenth- and Seventeenth-Century Europe*, Clarendon Press, 1985.

Jordanova, L. J., and Roy Porter (eds.), *Images of the Earth: Essays in the History of the Environmental Sciences*, 2nd ed., British Society for the History of Science, 1997.

Kölbl-Ebert, Martina (ed.), *Geology and Religion: A History of Harmony and Hostility*, Geological Society, 2009.

Krige, John, and Dominique Pestre (eds.), *Science in the Twentieth Century*, Harwood Academic, 1997.

LeGrand, H. E. *Drifting Continents and Shifting Theories: The Modern Revolution in Geology and Scientific Change*, Cambridge University Press, 1988.

Lewis, Cherry, *The Dating Game: One Man's Search for the Age of the Earth*, Cambridge University Press, 2000.

Lewis, Cherry, and S. J. Knell (eds.), *The Age of the Earth: From 4004 BC to AD 2002*, Geological Society, 2001.

Marty, Martin E., and R. Scott Appleby (eds.), *Fundamentalisms and Society*, University of Chicago Press, 1993.

Numbers, Ronald L., *The Creationists: From Scientific Creationism to Intelligent Design*, University of California Press, 2006 [first edition, 1993].

O'Connor, Ralph, *The Earth on Show: Fossils and the Poetics of Popular Science, 1802–1856*, University of Chicago Press, 2007.

Oldroyd, David R. (ed.), *The Earth Inside and Out: Some Major Contributions to Geology in the Twentieth Century*, Geological Society, 2002.

Oreskes, Naomi, *The Rejection of Continental Drift: Theory and Method in American Earth Science*, Oxford University Press, 1999.

Oreskes, Naomi (ed.), *Plate Tectonics: An Insider's History of the Modern Theory of the Earth*, Westview Press, 2001.

Poole, William, *The World Makers: Scientists of the Restoration and the Search for the Origins of the Earth*, Peter Lang, 2010.

Porter, Roy, *The Making of Geology: Earth Science in Britain, 1660–1815*, Cambridge University Press, 1977.

Rappaport, Rhoda, *When Geologists Were Historians, 1665–1750*, Cornell University Press, 1997.

Raup, David M., *The Nemesis Affair: A Story of the Death of Dinosaurs and the Ways of Science*, W. W. Norton, 1986.

Richet, Pascal, *A Natural History of Time*, University of Chicago Press, 2007 [*L'Age du Monde*, Éditions du Seuil, 1999].

Roger, Jacques, *Buffon: A Life in Natural History*, Cornell University Press, 1997 [*Buffon: Un Philosophe au Jardin du Roi*, Fayard, 1989].

Rossi, Paolo, *The Dark Abyss of Time: The History of the Earth and the History of Nations from Hooke to Vico*, University of Chicago Press, 1984 [*I Segni di Tempo*, Feltrinelli, 1979].

Rudwick, Martin J. S., *Bursting the Limits of Time: The Reconstruction of Geohistory in the Age of Revolution*, University of Chicago Press, 2004.

———, *Georges Cuvier, Fossil Bones, and Geological Catastrophes*, University of Chicago Press, 1997.

———, *The Great Devonian Controversy: The Shaping of Scientific Knowledge among Gentlemanly Specialists*, University of Chicago Press, 1985.

———, *Lyell and Darwin, Geologists: Studies in the Earth Sciences in the Age of Reform*, Ashgate, 2005.

———, *The Meaning of Fossils: Episodes in the History of Palaeontology*, 2nd ed., University of Chicago Press, 1985.

———, *The New Science of Geology: Studies in the Earth Sciences in the Age of Revolution*, Ashgate, 2004.

———, *Scenes from Deep Time: Early Pictorial Representations of the Prehistoric World*, University of Chicago Press, 1992.

———, *Worlds Before Adam: The Reconstruction of Geohistory in the Age of Reform*, University of Chicago Press, 2008.

Schneiderman, Jill S., and Warren D. Allmon (eds.), *For the Rock Record: Geologists on Intelligent Design*, University of California Press, 2010.

Schopf, J. William, *Cradle of Life: The Discovery of Earth's Earliest Fossils*, Princeton University Press, 1999.

Secord, James A. (ed.), *Charles Lyell: Principles of Geology*, Penguin Books, 1997.

Sepkoski, David, and Michael Ruse (eds.), *The Paleobiological Revolution: Essays on the Growth of Modern Paleontology*, University of Chicago Press, 2009.

Sepkoski, David, *Rereading the Fossil Record: The Growth of Paleobiology as an Evolutionary Discipline*, University of Chicago Press, 2012.

Van Riper, A. Bowdoin, *Men among the Mammoths: Victorian Science and the Discovery of Human Prehistory*, University of Chicago Press, 1993.

Wyse Jackson, Patrick, *The Chronologers' Quest: The Search for the Age of the Earth*, Cambridge University Press, 2006.

# SOURCES OF ILLUSTRATIONS

Fig. 3.4    Author's design.

Figs. 3.5, 3.6    Hutton, *Theory of the Earth*, 1795, vol. 1, pl. 3 and p. 200.

## Chapter 4

Figs. 4.1, 4.8    Knorr and Walch, *Merkwürdigkeiten der Natur*, vol. 1, 1755, Tab. XIIIb, Tab. XIa.

Fig. 4.2    Trebra, *Erfahrungen vom Innern der Gebirge*, 1785, Taf. VI.

Fig. 4.3    Arduino, MS section of Valle d'Agno, 1758, in Arduino archive, bs.760, IV.c.11, Biblioteca Civica di Verona.

Fig. 4.4    Hamilton, *Campi Phlegraei*, 1776, pl. 6.

Fig. 4.5    Faujas, *Volcans éteints*, 1778, pl. 10.

Fig. 4.6    Desmarest, "Détermination de trois époques" in *Mémoires de l'Institut National*, vol. 6, pl. 7.

Fig. 4.7    Lamanon, "Fossiles de Montmartre" in *Observations sur la Physique*, vol. 19 (1782), pl. 3.

Fig. 4.9    Hunter, "Observations on the bones near the River Ohio" in *Philosophical Transactions of the Royal Society*, vol. 58 (1769), pl. 4.

## Chapter 5

Fig. 5.1    Bru de Ramón, print in Cuvier archive 634(2), Bibliothèque Centrale, Muséum National d'Histoire Naturelle, Paris.

Fig. 5.2    Tilesius [Wilhelm von Tilenau], "De skeleto mammonteo Sibirico" in *Mémoires de l'Académie Impériale des Sciences de St Pétersbourg*, vol. 5 (1815), pl. 10.

Fig. 5.3    Cuvier, MS drawing, in Cuvier archive 635, Bibliothèque Centrale, Muséum National d'Histoire Naturelle, Paris.

Fig. 5.4    Cuvier, *Ossemens Fossiles* (1812), vol. 1, p. 3, author's translation.

Fig. 5.5    De la Beche, MS drawing, 1820, in De la Beche archive, MS 347, Department of Geology, National Museum of Wales, Cardiff.

Fig. 5.6    Hall, "Revolutions of the Earth's surface" in *Transactions of the Royal Society of Edinburgh*, vol. 7 (1814), pl. 9.

Figs. 5.7, 5.9    Buckland, *Reliquiae Diluvianae*, 1823, pls. 17, 21.

Fig. 5.8    Conybeare, "Hyaena's den at Kirkdale," lithographed print, 1823.

## Chapter 6

Figs. 6.1, 6.2    Cuvier, *Ossemens Fossiles*, 1812, vol. 1, part of "Carte géognostique" and pl. 2, fig. 1.

Fig. 6.3    Englefield, *Isle of Wight*, 1816, pl. 25.

Fig. 6.4    De la Beche, "*Duria antiquior*" print, 1830.

Figs. 6.5, 6.6   Buckland, *Geology and Mineralogy*, 1836, vol. 2, parts of pl. 1.

Fig. 6.7   Brongniart and Desmarest, *Crustacés Fossiles*, 1822, part of pl. 1.

Fig. 6.8   Goldfuss, *Petrifacta Germaniae*, vol. 3, 1844, frontispiece.

Fig. 6.9   De la Beche, *Researches in Theoretical Geology*, 1834, frontispiece.

### Chapter 7

Fig. 7.1   Mary Buckland to Whewell, 12 May 1833, MS letter in Whewell papers, a 66/31, Trinity College, Cambridge.

Fig. 7.2   [Rennie], *Conversations on Geology*, 1828, pls. [3, 5].

Fig. 7.3   Mantell, *Wonders of Geology*, 1838, frontispiece.

Fig. 7.4   Scrope, *Geology of Central France*, 1827, p. 165.

Fig. 7.5   Lyell, *Principles of Geology*, vol. 1, 1830, frontispiece.

Fig. 7.6   De la Beche, *Awful Changes* print [1830].

Figs. 7.7, 7.8   Rudwick, *Worlds Before Adam*, 2008, fig. 13.7 and fig. 35.3.

Fig. 7.9   Agassiz, *Études sur les Glaciers*, 1840, pl. 17.

### Chapter 8

Fig. 8.1   Geikie, *Great Ice Age*, 1894, plate XIV.

Fig. 8.2   Schmerling, *Recherches sur les Ossemens Fossiles*, 1833–34, vol. 1, plate I.

Fig. 8.3   Boucher de Perthes, *Antiquités Celtiques et Antédiluviennes*, vol. 1, 1847, reproduced in Donald K. Grayson, *Establishment of Human Antiquity*, Academic Press, 1983, p. 124.

Fig. 8.4   Prestwich, "Exploration of Brixham Cave," *Philosophical Transactions of the Royal Society*, vol. 163, 1873, p. 550.

Fig. 8.5   Agassiz, *Recherches sur les Poissons Fossiles*, 1833–43, vol. 1, plate at p. 170.

Fig. 8.6   Gaudry, *Animaux Fossiles et Géologie de l'Attique*, vol. 2, 1867, p. 354.

Fig. 8.7   S. J. Mackie, "Aeronauts of the Solenhofen Age," *Geologist*, vol. 6, 1863, plate I.

Fig. 8.8   Boitard, "L'Homme Fossile," *Magasin Universel*, vol. 5, 1838, p. 209.

### Chapter 9

Fig. 9.1   Smith, *Chaldean Account of Genesis*, 1876, p. 10.

Fig. 9.2   Hawkins, "Visual education as applied to geology," *Journal of the Society of Arts*, vol. 2, 1854, p. 446.

Fig. 9.3   Prévost, "Formation des terrains," in *Candidature de Prévost*, 1835, plate.

Fig. 9.4   Lugéon, "Grandes Nappes de Recouvrement des Alpes," *Bulletin de la Société Géologique de France*, ser. 4, vol. 1, 1902, fig. 3, p. 731.

Fig. 9.5    Bertrand, "Châine des Alpes," *Bulletin de la Société Géologique de France*, ser. 3, vol. 15, 1887, p. 442, fig. 5.

Figs. 9.6, 9.9    Phillips, *Life on the Earth*, 1860, pp. 66, 51.

Fig. 9.7    Barrande, *Système Silurien du Centre de la Bohême*, vol. 1, 1852, pl. 10.

Fig. 9.8    Dawson, *Life's Dawn on Earth*, 1875, pl. IV.

## Chapter 10

Fig. 10.1    Zeuner, *Dating the Past: An Introduction to Geochronology*, Methuen, 1946, fig. 17, p. 51.

Fig. 10.2    Wegener, *Entstehung der Kontinente und Ozeane*, 1922, fig. 2, p. 5.

Fig. 10.3    Köppen and Wegener, *Klimate der geologischen Vorzeit*, Borntraeger, 1924, fig. 3, p. 22.

Fig. 10.4    Du Toit, *Our Wandering Continents: An Hypothesis of Continental Drifting*, Oliver & Boyd, 1937, fig. 9, p. 76.

Fig. 10.5    Holmes, "Radioactivity and Earth Movements," *Transactions of the Geological Society of Glasgow*, vol. 18, 1931, figs. 2, 3, p. 579.

Fig. 10.6    J. E. Everett and A. G. Smith, "Genesis of a geophysical icon . . . ," *Earth Sciences History*, vol. 27 (2008), p. 7, fig. 5, reproduced from E. Bullard, J. E. Everett, and A. G. Smith, "The fit of the continents . . . ," *Philosophical Transactions of the Royal Society of London*, vol. A 258 (1965), pp. 41–51, fig. 8.

Fig. 10.7    Naomi Oreskes, *Plate Tectonics*, Westview, 2001, p. 48, reproduced from A. D. Raff and R. G. Mason, "Magnetic survey . . . ," *Bulletin of the Geological Society of America*, vol. 72 (1961), pp. 1267–70.

Fig. 10.8    Xavier Le Pichon, Jean Francheteau, and Jean Bonnin, *Plate Tectonics*, Elsevier, 1973, fig. 27, p. 83.

Fig. 10.9    Anthony Hallam, *A Revolution in the Earth Sciences: From Continental Drift to Plate Tectonics*, Clarendon, 1973, fig. 34, p. 79.

## Chapter 11

Fig. 11.1    Mary Leakey and Richard Hay, "Pliocene footprints in Lateolil beds at Lateoli, northern Tanzania," *Nature*, vol. 278, 22 March 1979, fig. 7, p. 322.

Fig. 11.2    Whittington, *The Burgess Shale*, Yale University Press, 1985, fig. 4.70.

Figs. 11.3, 11.6    Preston Cloud, *Oasis in Space: Earth History from the Beginning*, Norton, 1988, fig. 10.9 A, p. 239; and fig. 11.5 A, p. 262.

Fig. 11.4    Glaessner, "Pre-Cambrian animals," *Scientific American*, vol. 204, 1961, p. 74.

Fig. 11.5    Harland and Rudwick, "The Great Infra-Cambrian Ice Age," *Scientific American*, vol. 212, 1964, p. 30.

Fig. 11.7    Arizona meteor crater, Wikimedia Commons.

Fig. 11.8    C. S. Beals, M. J. S. Innes, and J. A. Rothenberg, "Fossil meteorite craters," fig. 1, in Barbara M. Middlehurst and Gerard Peter Kuiper, *The Moon, Meteorites and Comets*, University of Chicago Press, 1963, p. 237.

Fig. 11.9    "The Blue Marble," Wikimedia Commons.

## Chapter 12

Fig. 12.1    "A man-made world," *The Economist*, 28 May 2011, p. 81.

Fig. 12.2    Unger, *Die Urwelt*, 1851, Atlas, Taf. 9.

# ACKNOWLEDGMENTS

This book reflects the shape of my whole second career as a historian (with a few allusions to my first as a scientist), so it is hardly possible to record all the colleagues to whom I am indebted: those at Cambridge who first helped me to learn how to think as a historian of the sciences; and later those in several countries around the world, whose live discussions and published research have been an invaluable stimulus for my own work. A second kind of debt is to my students, first at Cambridge, later in Amsterdam, Princeton, and San Diego, and briefly in Utrecht, on whom I tried out successive versions of the narrative and analysis embodied in this book. And since few of them were planning to become professional historians of the sciences, it was realistic to treat them all as a sample of the intelligent general readers whom I hope will find this book of interest. Another such sample are those of my friends, most of them not in the academic world, who have generously given their time to read one or more draft chapters of the book and to tell me whether they found the prose and the pictures accessible and interesting. Since some of them want to remain anonymous I am extending that anonymity to them all; but I hope they already know how much I have valued their comments, and their encouragement to complete the book at the same level, without dumbing-down. Finally, I am hugely indebted to the editors, designers, and others at the University of Chicago Press, who over many years have made my books—quite apart from their content—attractive to handle and read; and particularly to Karen Darling, who has overseen the production of this my last book not only with deep professionalism and insight but also with unfailing courtesy and thoughtfulness.

# INDEX

Page numbers in italics refer to figures and their captions.